Water Interactions – A Systemic View

Why we Need to Comprehend the Water–Climate–Energy–Food–Economics–Lifestyle Connections

Water Interactions – A Systemic View

Why we Need to Comprehend the Water–Climate–Energy–Food–Economics–Lifestyle Connections

Gustaf Olsson

Published by

IWA Publishing
Unit 104–105, Export Building
1 Clove Crescent
London E14 2BA, UK
Telephone: +44 (0)20 7654 5500
Fax: +44 (0)20 7654 5555
Email: publications@iwap.co.uk
Web: www.iwapublishing.com

First published 2022
© 2022 IWA Publishing

Disclaimer
The information provided and the opinions given in this publication are not necessarily those of IWA and should not be acted upon without independent consideration and professional advice. IWA and the Editors and Authors will not accept responsibility for any loss or damage suffered by any person acting or refraining from acting upon any material contained in this publication.

British Library Cataloguing in Publication Data
A CIP catalogue record for this book is available from the British Library

ISBN: 9781789062892 (Paperback)
ISBN: 9781789062908 (eBook)
ISBN: 9781789062915 (ePub)

This eBook was made Open Access in August 2022.

© 2022 The Author.

Contents

Chapter 6
The energy perspective **85**

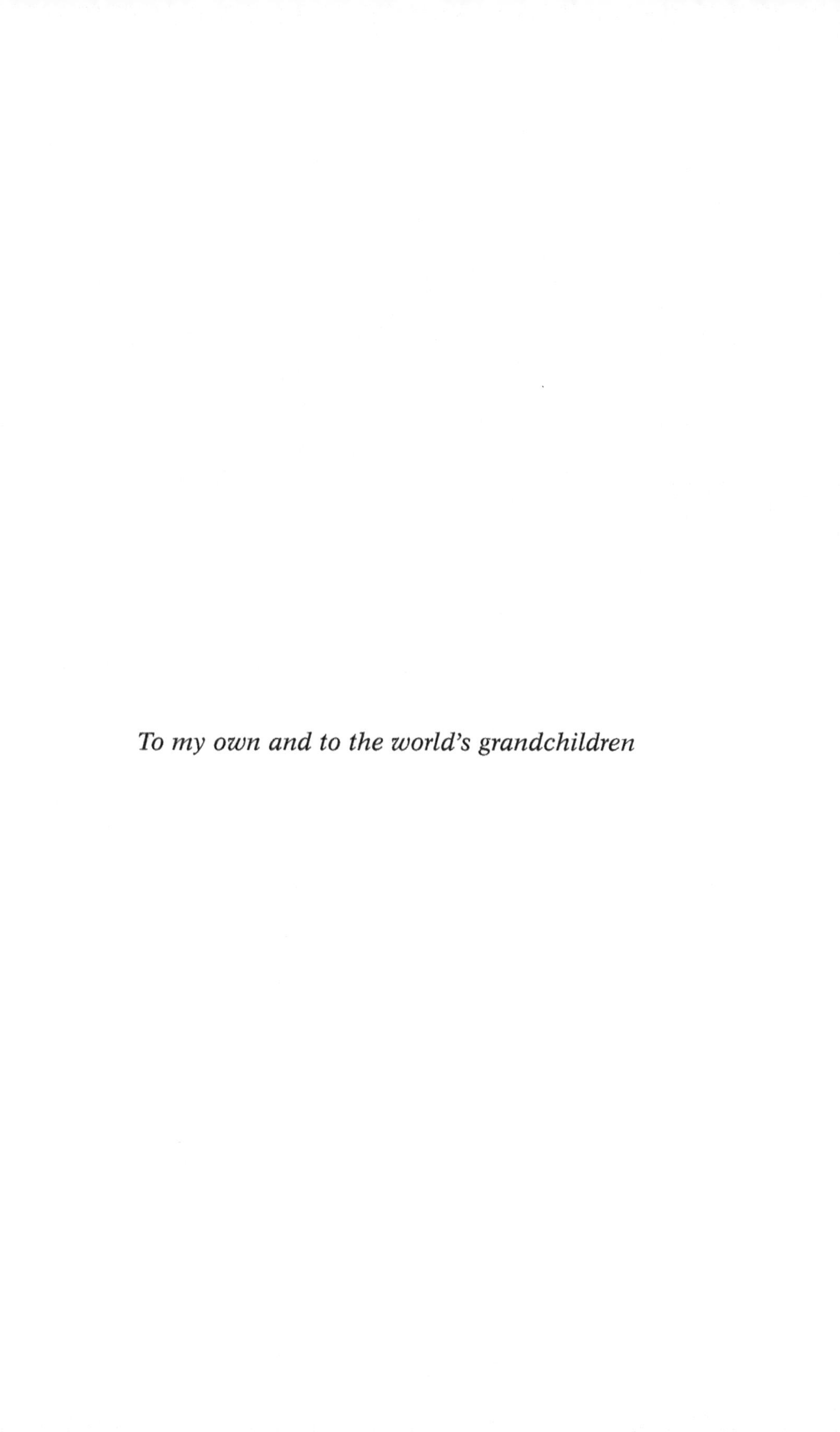

To my own and to the world's grandchildren

About the author

Gustaf Olsson is a professor in industrial automation and since 2006 professor emeritus at Lund University, Sweden. His has devoted his research to control and automation of urban water systems, power production, electrical power systems, and industrial processes. Since his retirement, he has focused on the water–energy nexus: how energy exploration, generation, and use are related to water availability and quality, and how water operations and consumption are dependent on energy. Naturally, couplings to climate change, population increase, and urbanization are crucial.

He was part-time guest professor for more than ten years at Tsinghua University, Beijing, from 2006, and at the Technical University of Malaysia. He is an honorary faculty member at the University of Exeter, UK, and advisor to several international research groups and programs. He has been the editor-in-chief of *Water Science and Technology* and member of the IWA (International Water Association) Board of Directors. In 2010 he received the IWA Publication Award. In 2012 he was the awardee of an honorary membership of IWA as well as an honorary doctor degree at the Technical University of Malaysia. In 2014 he was appointed Distinguished Fellow of the IWA. For almost a decade he has been mentoring IWA Young Water Professionals from all continents.

Gustaf has authored 12 international books (some of them translated into Russian, German, Korean, Chinese and the Persian language), contributed chapters to more than 20 books and published over 200 scientific papers. His book *Water and Energy* first appeared in 2012 followed by a 2nd edition in 2015. With Pernille Ingildsen, he co-authored *Smart Water Utilities: Complexity Made Simple* (IWA Publishing 2016, now open access). His book *Clean Water using Solar and Wind: Outside the Power Grid* was published in 2018 (IWA Publishing, now open access). He has been invited to several countries and international conferences to present results around the topics of water and energy.

doi: 10.2166/9781789062908_xv

Preface

The desire to understand was the key driving force to write this book. The world is a complex place with a hodgepodge of interacting problems and possibilities. We are facing complex problems like global warming, climate change, water scarcity, droughts, floods, food crises, energy limitations, air pollution, and absurd economic inequalities between nations and people. Complex problems cannot be solved by looking at one component at a time. Most incidents are connected to others. In our world of specialization, solutions require cooperation, mutual respect, and understanding what to expect from other specialists.

The aim of this book is to demonstrate and describe how climate change, water, energy, food, economics, and lifestyle are intimately interconnected. We – all of us in the global village – depend on each other. This has become so apparent during the pandemic from 2020 onwards. Global warming and its impact on climate will affect every continent and nation and there is no vaccine. We must understand these interconnections and act accordingly.

Deniers do not see the connections or do not wish to understand: rain has fallen for the first time on record at the Greenland Summit Station, the highest point on the Greenland Ice Sheet (3216 metres). On 14 August, 2021, the temperature remained above freezing for about nine hours. The warm air caused extreme rain on the ice sheet – the heaviest rainfall there since record keeping began in 1950, according to the National Snow and Ice Data Center at the University of Colorado. Rain in Iceland is another anomaly where deniers do not recognize projections, couplings to climate or further explanations; the rain can be considered just a fun fact. Isolated from fires in Siberia, floods in Germany, glacial melt, calving of icebergs the size of states, hurricane seasons getting longer and longer, outgassing of methane in the tundra, droughts, overwhelming cold snaps, coral bleaching, fires in the rain forests, and then a little strange rain would be nothing but odd weather. So long as all these events are separated from one another, they can all be explained away. But taken as a whole, they can only be attributed to a fundamental change on our planet.

Over the years there have been several proposals for specific design solutions, control methods, or operational guides in terms of green technology, water and energy efficiency, food production, geoengineering methods, and space settlements on Mars to solve the environmental and resource problems of a finite Earth. Recalling the famous phrase attributed to Einstein, *'we cannot solve our problems with the same thinking we used when we created them,'* we must become aware of the close couplings. An integrated approach to face the complex challenges is necessary. This includes not only innovative technical solutions but several non-technical issues too. It's equally important to address political, organizational, and economic topics. *Changes in our attitudes and lifestyles are crucial* if we wish to create a more sustainable use of natural resources. Saving one MW ('negawatt') is cheaper than producing one. There is a sense of urgency when we see dwindling water resources in many places and when the increasing use of energy will further aggravate climate change.

Our situation today is not an accident – it is the consequence of the way we think and act. We have failed for decades to act, having listened to scientific evidence of climate change and its causes. It is crucial when getting to know the context, to see the connections, listening not only to lobbyists but to people that are affected by the changes. The Covid-19 pandemic has taught us that we have an interconnectedness and mutual responsibility towards one another and towards our Earth. Caring for our fellow citizens and caring for our environment are the same thing. As expressed by Chief Oren Lyons (Native American Faithkeeper of the Turtle Clan, member of the Indigenous Peoples of the Human Rights Commission of the UN, and professor of American Studies at University of Buffalo, New York): *'Man sometimes thinks he's been elevated to be the controller, the ruler, but he's not. He's only part of the whole. Man's job is not to exploit, but to oversee, to be a steward. Man has responsibility, not power.'*

We will describe several examples of our exploitation of nature and our self-destructive behaviour. Again, Oren has articulated: *'The law says if you poison the water, you'll die. The law says that if you poison the air, you'll suffer. The law says if you degrade where you live, you'll suffer ... If you don't learn that, you can only suffer. There's no discussion with this law.'*

It has been said many times that we have to listen to science to understand dangers and connections. However, we also must look for and recognize dangers that are not described by natural sciences: greed, apathy, and selfishness. Science cannot provide a moral compass for our actions. This must be achieved by cultural and spiritual transformations and then we have to recognize how our economic systems and our lifestyle are intimately connected to climate, water, energy, and food.

The *mission of the book* is to describe how various objects are related to each other. Unless we understand these couplings, we cannot even start solving the complex problems we face. Various solutions or approaches to solutions presented here will probably not provide any innovative new answers to the water, energy, food, and economy challenges that we are facing. This herculean task is far too unrealistic for an individual. However, understanding the couplings is an important beginning.

The IPCC Working Group II contribution to the Sixth Assessment Report was released only a few weeks before the completion of this manuscript. It adds further evidence of the enormous threat that the world is facing from climate change but does not change any of the conclusions and statements made in this book. Our attention concerning the current climate crisis and resource challenges must not be pushed into the background because of the Russian invasion of Ukraine. There is a rapidly closing window for climate change action.

Vrångö, Swedish west coast archipelago
March 2022
G.O.

Acknowledgements

For decades I have had the privilege to learn from so many people. They have encouraged me, corrected me, and widened my perspective. I have learnt a lot by guiding 23 PhD and hundreds of MSc students in Sweden, and from several other PhD students while a guest professor at Tsinghua University, Beijing, China; Technical University of Malaysia; and at universities in Johannesburg and Cape Town, South Africa.

Allan R. Hoffman, former Senior Analyst, US Department of Energy, was a most influential source of knowledge when I started studying the water–energy nexus. We have remained in contact for more than a decade, and he has generously shared his knowledge and offered me feedback on my ideas and given me new perspectives.

Gianguido Piani was already my co-author more than 30 years ago. He is one of these exceptional renaissance style people with a wide knowledge, not only in technology, but in humanities and language. Having lived in many countries working as an energy and automation specialist, he has given me plenty of new outlooks and a lot of constructive feedback.

Lars Gertmar, my professor colleague at Lund University, Sweden, since the early 1990s, has generously spent a lot of effort to review, correct, and challenge me, given his extensive experience in all kinds of aspects of energy transition, from both industrial and academic environments.

Rune Andersson, professor emeritus in global health and infectious diseases, University of Gothenburg (Göteborg, Sweden), has kindly given me feedback on all the health aspects discussed in the book.

Cello Vitasovic, Los Angeles, USA and Pernille Ingildsen, Denmark, have been my close collaborators for a long time. I first met Cello when he was a PhD student more than 40 years ago. Pernille was my PhD student more than 20 years ago. Our cooperation has continued over the years and resulted in several publications. Both Cello and Pernille have given me a lot of inspiration, joy, new perspectives, and courage to proceed with this work, and have through the writing project offered me many good ideas on how to make the content more relevant.

Lawrence Jones, Vice President International Programs, Edison Electric Institute, Washington DC, USA, and I have had regular contacts for more than two decades. Lawrence has a true global perspective and knowledge that is a key prerequisite to solve complex problems. His experience and feedback have been invaluable to me in my writing.

Francisco J. Márquez-Fernández, Lund University, has been a great support, given his electrical engineering background. We have penetrated many issues related to the book in our discussions.

Nenibarini Zabbey, University of Port Harcourt, Nigeria, is not only a great colleague since a decade. He has given me a lot of insight in the costs of oil exploration. He is a courageous role model for water professionals.

Mia Engstad and Mikael Engstad in Sweden made me aware of and demonstrated the potential of regenerative farming. This motivated me to study this crucial topic and gave me new perspectives on food production.

Jing Liu has managed to use his scientific knowledge in biotechnology and automation in combination with his leadership of a successful company (BPC Instruments, Lund, Sweden). I have learnt a lot from him during two decades of cooperation.

Margareta Ingelstam, Swedish author and educator, is a pioneer and veteran in the work on peace and nonviolence and a great role model. During the past year I have had the fortune to work closely together with her to explore how experiences from the peacebuilding movement can be applied to education and empowerment on Climate Justice and Climate Security.

And all of you that helped with your conversations, ideas, and encouragement (in alphabetic order): Per Duregård, Lars Fränne, Ulf Jeppsson, Leif Nilsson, Lars Odén, Lars Gunnar Sundin (all from Sweden), Zaini Ujang (Malaysia), and Peter Vanrolleghem (Canada).

I had no plan to write a book like this when Mark Hammond, Books Commissioning Editor at IWA Publishing, in 2021 encouraged me to write some extension of my previous books *Water and Energy* and *Clean Water Using Solar and Wind*. One probably must be foolhardy to accept, but I could not resist the temptation to try. I wish to thank the cooperative and friendly staff at IWAP.

Kirsti, my patient, incomparable wife, miraculously loves me despite my odd habits of work. Regularly I ask her advice how to describe various phenomena concerning climate, water, energy, and lifestyle. Without her support and understanding I could not have finished this work. And more essential, she provides me with a more complete perspective of life.

And you, my children and grandchildren, you have motivated me to write the book. I tried my best, thinking of your future.

Prologue – Some experiences in my lifetime

What has been will be again,
what has been done will be done again;
there is nothing new under the sun. Ecclesiastes 1:9

The awareness that Mother Earth has limited resources is not new. During my life I have gradually learnt how nature, resources, industrial development, and lifestyle are interconnected. I have discovered that at the age of 80+ years I see events and experiences in life in context in a way that I didn't when I was younger. This prologue is a subjective narrative from my lifetime and how I experienced pathways toward today concerning the state of water, energy, climate, population, and food. My aim is to set the scene of today, providing some background information for the rest of the book.

I have lived a privileged life. My family was not wealthy, a typical middle-class Swedish family. Still, compared to most people in this world we have been fortunate. I was born in 1940 in Sweden, in the second year of World War II. Finland was fighting the Soviet Union; some 70 000 Finnish children were brought to Sweden to escape the war (my wife was one of them). Germany had invaded Norway and Denmark, while Sweden managed to escape getting involved in the war. The German offence in Western Europe continued. The Battle of Britain was initiated by Hitler. A Swedish textbook on the German language had the text *'Der Führer hat Ordnung geschaffen'* (The leader Adolf Hitler has created order). Towards the end of the 1940s the text was changed to *'Gott hat die Welt geschaffen'* (God has created the world).

1940 global population: 2.4 billion
Global CO_2 concentration: $\cong 280^* + 20 = 300$ ppm
(parts per million or mg/l)
* = the pre-industrial level

There was a great shortage of gasoline in Sweden during the war and towards the end of the year 1940 all private consumption of gasoline for cars was forbidden. The cars had to be fuelled with wood gas. Towards the end of the war in 1945 I was sufficiently old to remember the wood gas container mounted on the back of our car. There was a special smell of burning and my father had some dirty gloves to handle the container. Fossil fuel consumption was dominated by coal, and our apartment building was heated by coal for the whole decade. I recall how coal was delivered to our building in sacks made of jute fabric. The delivery men were really dirty. Later, when coal was replaced by oil it was praised as a truly clean solution for heating. I was reminded of my childhood when I was working at Tsinghua University in Beijing after 2006. Then the university was still heated by coal. However, all coal-powered energy now has been replaced by natural gas in Beijing.

Many years after my childhood, I studied the geopolitical importance of oil and particularly how energy and oil capacity was a decisive factor in World War II operations. Trucks, tanks, cars, and airplanes needed gasoline. Refined oil was used for making TNT in bombs and synthetic rubber for tyres as well as a lubricant for all kinds of machinery. The American oil industry played a remarkable role by supplying the Allied forces with oil. During the 1930s the USA had provided 50% of the global oil supply. Oil tankers had been easy targets for German submarines between the Mexican Gulf and the US East Coast. In the early years of World War II two pipelines were built from Texas to the East Coast. They were protected from enemy attack. Oil from the USA provided some 85% of all the oil – around 1.1 billion (10^9) m^3 – used by the Allied Nations during World War II. An apparent reason for the German Army operations in northern Africa led by General Erwin Rommel was to capture the Suez Canal. This would ensure the route to the oil fields of the Middle East. A major driving force for Hitler to invade the Soviet Union in 1941 was to capture the Soviet oil fields in the Caucasus Mountains. This would have given Germany the necessary oil to fuel the German war machine. The German Field-Marshall Karl Gerd Von Rundstedt admitted after the war that German deficiency in oil was a key reason for losing the war.

Also, the War in the Pacific was fundamentally a war over oil resources. Japan lacked its own oil supplies. The USA had been alarmed by Japanese aggression in Asia and all US oil shipments to Japan were ordered to cease in August 1941, four months before Pearl Harbour. Japan invaded colonial Indonesia in March 1942, occupying the Dutch oil fields. For the rest of the war American submarines waged unrestricted attacks against Japanese oil tankers in the Pacific.[1] Today, as we expand solar and wind, the geopolitical importance of oil ought to decrease.

Even if Sweden was spared from the war there were several consumer restrictions. Private citizens had ration stamps for meat, coffee, sugar, and many other food products. I can vividly recollect my mother telling us to appreciate some special food.

Before I was 5 years old we once travelled some 300 km to the west coast of Sweden. My father was in the military protecting the Swedish border and my mother brought my two elder brothers and me for a short vacation at the seaside. Since travelling was limited also after the end of the war this was felt like an exclusive experience.

In 1948 the family had purchased a little house for summer vacations, located only 8 km away from our town apartment. This meant we all could use bicycles, even if I was only 7 years old. I have clear memories related to water, energy and food. Our town had no biological wastewater treatment plant, and most of the untreated effluent water was flowing into the lake Hjälmaren where we used to swim. My older brother developed serious eczema caused by the polluted water and could not swim for the whole summer. We were reminded all the time to turn off the light when we left any room. Beef was a luxury that we sometimes could get on Sundays.

Since my childhood there has been an incredible development in mobility, which is also reflected in my own life. The family trips until I was almost 15 years were limited to one or two 200 km trips per year to my grandparents. A good friend in my hometown won a competition and earned a trip to Brazil in 1955. This got the attention of the local newspaper and the sensational information that she should bring swimwear in the Swedish winter! My first visit outside Scandinavia took place in 1964, during my fourth year at the Royal Institute of Technology, when we made a study tour by train to Paris and Geneva. Charter flights made travelling by air affordable and in the 1960s we could fly to the Canary Islands in 8 hours in propeller-driven airplanes. Since 1970 I made at least one intercontinental flight every year until I had around 200 flight hours in a year. This made me realize that I was part of the climate problem. Now I have minimized flying, and every trip has to be highly motivated from a climate point of view.

By 1950 I can recall the grown-ups talking about the industrial development after the war. Smokestacks became symbols of development. However, we became aware of air pollution quite early. Close to my hometown an oil shale industry had been established during the war since all import of oil was restricted. The furnaces in Kvarntorp produced as much as 100 000 m^3 of oil annually. We could smell the air pollution and we saw the slag heap growing. The ash was hot when it was dumped on the heap, some 600°C, so the burning hill emitted a very particular smell. This childhood memory gave me some understanding of pollution caused by fossil fuel extraction. The hill is more than 100 metres high and is now an attraction, being the highest point in the county of Närke in mid-Sweden.

1950 global population: 2.54 billion.
Global CO$_2$ concentration: $\cong 280 + 28 = 308$ ppm

During the 1950s the cold war influenced a lot of discussions and fear of a nuclear war was obvious. However, in 1955 a landmark conference was organized by UN, the 'Conference on the Peaceful Uses of Atomic Energy'. This Geneva meeting was the largest scientific meeting ever held, with an estimated 25 000 participants. For me personally it determined my early career. Nuclear power was apparently going to save the world and 9 years later I got my first employment at the Swedish State Power Board (now Vattenfall) as a nuclear engineer.

I comprehended the geopolitical power of oil for the first time in 1956. The Suez crisis started in October, when Israeli armed forces pushed into Egypt toward the Suez Canal after Egyptian president Nasser had nationalized the Canal that controlled the import of two thirds of the European oil. This created a gasoline crisis in Sweden and elsewhere. No private cars were allowed to be used from Saturday night until Monday (Saturday were working days at that time). Our teenager group enjoyed the car-free weekends and the opportunity to walk anywhere in the middle of the main road.

Alarming accidents in Japan got the world's attention in 1956. A lot of people had been seriously sick and even died due to severe mercury poisoning. For years we had noticed with admiration how Japan had raised itself from the catastrophe of the war and developed industrial production at a rate never seen before. Suddenly this picture became stained. Years later we understood the reasons behind and the chain of processes that triggered the tragedy. The diseases and deaths were caused by methylmercury – an extremely toxic organometallic cation $[CH_3Hg]^+$ – in the industrial wastewater from a chemical factory in Minamata, located on the west coast of the southern Japanese island of Kyushu. The plant had been in operation since 1932 and continued its production until 1968. Some of the mercury sulfate in the wastewater bioaccumulated and biomagnified in shellfish and fish in the Bay and the sea. The local population ate the seafood and got mercury poisoning. Almost 2000 people had died, and several thousand people were the victims. For more than a decade the authorities neglected the problems and did little to prevent the epidemic. The Minamata disaster was a wake-up call that reckless industrial production has a huge price, most often not paid by the industry owners or the customers.[2]

By the end of the 1950s we were experiencing great optimism about the future. I have a vivid memory of 4 October, 1957, when the Soviet Union launched Sputnik. This was the inauguration of the Space Age, and for a 16-year-old boy this was the realization of science fiction. The satellite circled Earth once every hour and 36 minutes. It was clearly visible before sunrise and after sunset. Our local newspaper published a daily update of the times when Sputnik would be visible. The park near our home was packed with people waiting to see Sputnik. My mother did not believe that man should be in space, so she considered the whole spectacle as fake, at least for the first few weeks. Sputnik initiated the space race, and the USA were caught off-guard by the Soviet technological achievement. John F. Kennedy was elected president in 1960 and became the great hero among young people. Moreover, he declared to 'send a man safely to moon before 1970'.

1960 global population: 3.0 billion
Global CO_2 concentration: $\cong 280 + 36 = 316$ ppm (the first CO_2
measurements were initiated by Charles David Keeling at Mauna Loa
Observatory, Hawaii, March, 1958)

The 1960s was also a decade when Earth's limitations were realized and debated.

I have childhood memories of how we used DDT at home to spray in the closets to get rid of the common clothes moth. The memory of the smell is distinct. It was considered an efficient way to protect our clothes. In 1962 Rachel Carson published her revolutionary book *Silent Spring*. It documented the indiscriminate use of pesticides and accused the chemical industry of spreading disinformation, and public officials of accepting the industry's marketing claims unquestioningly. There are several similarities between this discussion, the dispute about the tobacco industry, and today's debate around climate change and the fossil fuel industry.

In Sweden, much less known internationally, the author Elin Wägner (1882–1949) had published the book *Fred med jorden* (Peace with the Earth) already in 1940, during World War II. She wrote about 'the demands from health and conscience' and criticized the exploration of nature, agricultural land, and human resources, long before ecology and environment were recognized by the general public. Since the World War I she had shown in more than 30 books and numerous articles that three views are closely related: gender equality, peace, and our environment. They depend on each other. Inequality causes war, war destroys human beings, and the environment is ruined if we believe that we can dominate and exploit nature.

In 1965 the Sholapur district in Maharashtra, India, suffered a severe drought having had no rain for 3 years, forcing people to move, causing a crisis in food production, and emergency slaughter of cattle. The Maharashtra Government contacted the Swedish Covenant Church (Svenska Missionsförbundet) to ask for help. A cooperative effort was initiated, called '1000 wells for India' where the Swedes collected money in their churches and sent volunteers to help. One of the pioneers was Oscar Carlsson together with his wife Ingrid, working in education in India.

Oscar became the leader of a foundation called Sholapur Well Service (SWS). On top of the well digging and drilling the little workshop manufactured hand pumps. Oscar was also an inventor and gradually developed a better pump design, the SWS pump, later pictured on an Indian stamp with the text Safe Water. Well number 1000 was completed in 1974. Lead by the Hindustani Covenant Church more than 3000 wells had been completed in 1986. The pump was recognized by UNICEF as a most reliable and cost-effective hand pump, called India Mark II. More than 3 million pumps have been produced and have been used all over India and in more than 50 other countries. Oscar never filed for a patent but wanted the pump to be freely available to anybody in need. He was the first individual to be awarded 'The Swede of the year in the World' in 1968. I had the privilege getting to know Oscar. In 1981 I got the opportunity to visit Sholapur - invited by another Swedish couple, Eva and Georg Smedberg, working unselfishly for the non-privileged people – to attend the ceremony in a village where a new pump had been installed. Hundreds of villagers had gathered around the well. To see the first flow of water coming out from the hand pump is one of my most unforgettable moments.

Of course, electric pumps made it possible for so many more people to get access to water. However, replacing hand pumps with electric pumps has created a new problem of overuse.

The global annual population growth rate had increased from around 1% after World War II to around 1.8% in the early 1960s. This naturally caused a lot of warnings about the challenge of feeding a large population. Paul R. Ehrlich, an American biologist (with the speciality of entomology, the study of insects) warned about the consequences of population growth and limited resources. In his early career he had been inspired by *Road to Survival*, by William Vogt, published in 1948. This was an early warning of the dangers of overpopulation. Ehrlich published his controversial book *The Population Bomb*[3] in 1968. The first sentence defines the tone: '*The battle to feed all of humanity is over.*' He suggested that various population control methods should be used, such as including 'various forms of coercion' such as eliminating 'tax benefits for having additional children,' to be used if voluntary methods were to fail.

In the decades after Ehrlich's alarm there has been a decline in the *growth* rate, as shown in Figure 0.1. The *fertility* rate for a zero-growth population is 2.1% and has decreased from 5.0% in 1950 to 2.4% in 2021. During the last 5 years the decline has been 0.4% per year. The UN predicts that the population growth rate will decline towards zero at the end of this century.

Georg A. Borgström (1912–1990), a Swedish scientist, geographer, and ecologist, devoted his research to the global hunger problem. He was a professor in plant physiology and an international authority on food production. During the 1960s and 1970s he published several books on food resources, with titles like *The Hungry Planet, Food for Billions, The Food and People Dilemma*, and other studies on the Earth's ecological limitations. Maybe the most famous book is *Gränser för vår tillvaro* (Limits to our Existence). He also described 'the great unfair state' as Europeans seized rich agricultural lands in America and Africa and were fishing all over the oceans. He claimed that Europe could get sufficient food only because of these 'ghost land areas'. Europe exported cereal products but could import protein-rich food. His ideas are still convincing.

During the following decades the green revolution took place, and the use of fertilizers and pesticides increased significantly. Huge irrigation projects increased food production, and as a result the food crisis has developed into

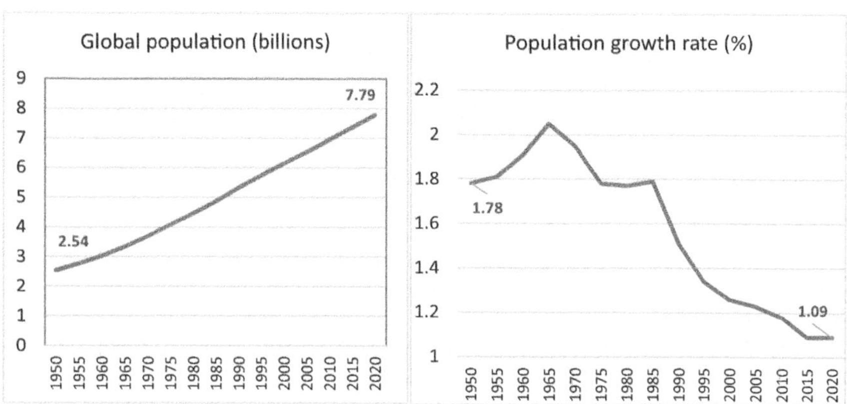

Figure 0.1 World population development since 1960.[4] The growth rate (5-year averages in the diagram) reached its maximum of 2.1% in 1968 and is down to 1.03% in 2021, the same level as before 1940.

a water resource crisis, where more than 70% of the global water use is for food production. However, the water crisis in fact had been a central theme in Borgström's early publications. He also noted that *'to increase the food production has demanded much more water than earlier normal harvests. A large part of the increased production has been bought with water that otherwise would replenish the groundwater. In Western Europe the increased harvests since early 1900 correspond to some 30–50 mm annual precipitation. Thus, the increasing harvests mean serious plundering of water reserves'* [my translation from Swedish]. Borgström also emphasized that those who recommend fast industrialization of the low-income countries often ignore the fact that this requires water availability beyond available limits. He stressed the fact that irrigation will disturb the natural water balance. Often irrigated land can be destroyed alarmingly quickly by salt formation, and he showed cases from places as different as Utah and Iraq.

It is not a new discovery that water reservoirs are causing health problems. Large water basins are sources for certain diseases, carefully described by Borgström in the 1960s. Malaria is the most well-known example. The malaria mosquitoes had already developed resistance to the pesticides being developed. An even larger threat came from the parasite disease bilharzia. In the 1960s some two thirds of the population in the Nile Delta were infected. The eggs are distributed in water and need various kinds of snails as hosts to hatch the larva. Before the Aswan dam was built there was an annual dry period that eliminated a large portion of the snails. Permanent irrigation made the snails thrive. So Borgström's early key message was to combine the technical innovations and developments with a biological framework.

The cost for oil transportation became apparent for us in 1967. The tanker *Torrey Canyon* – a huge ship of almost 300 metres length – ran aground off the

Cornwall west coast in the UK, spilling an estimated 94 000–164 000 m^3 of crude oil. At the time *Torrey Canyon* was the world's worst oil spill and it got a lot of media attention. This was the first supertanker accident, and it raised awareness of the risks of oil. To limit the consequence of the oil spill it was decided to set the wreck on fire by means of air strikes until the ship would sink. To achieve this, some 1500 tons of napalm and 45 m^3 of kerosene were used.[5]

Unfortunately, more accidents followed. In 1978 the *Amoco Cadiz* sank near the northwest coast of France and spilled some 260 000 m^3 of crude oil. This became the largest oil spill from an oil tanker in history.[6] Our oil dependence remains dangerous.

On 21 December, 1968, Apollo 8, the second Apollo mission, left Earth with the purpose to circle around the moon. Nobody had seen the back of our moon before and now we could see it during Christmas. And we saw the Earth appear behind the horizon of the moon. This picture, more than anything else before, symbolized our precious, vulnerable, fragile home, our only home.

Maybe the moon landing on 20 July 1969 will be one of the few memories of the 20th century. This was the highlight of the NASA Apollo programme and the fulfilment of President Kennedy's promise in 1961. To achieve this, Congress had to increase the NASA budget 20-fold. The price to pay troubled Kennedy when in a speech at Rice University in September 1962 he tried to motivate his citizens to make sacrifices: '*We choose to go to the Moon in this decade and do the other things, not because they are easy, but because they are hard; because that goal will serve to organize and measure the best of our energies and skills, because that challenge is one that we are willing to accept, one we are unwilling to postpone, and one we intend to win*'....., and his famous '*Ask not what your country can do for you ....*'. Maybe this should be a guideline for our decision makers when addressing the global warming challenge.

I spent the academic year 1970–71 as a postdoc at University of California, Los Angeles (UCLA) at the Systems Science Department. It gave me a deeper insight into systems thinking. A system is more than its parts or components. Interactions rather than parts in isolation must be considered. I also understood better how the same type of thinking can be used not only in technical processes but also in areas like astronomy, economics, sociology, and medicine. This insight has inspired me in my attempts to understand the nexus that is the topic of this book.

1970 global population: 3.7 billion
Global CO$_2$ concentration: $\cong 280 + 45 = 325$ ppm

From a personal point of view the year in Los Angeles presented a lot of sobering experiences. Flying in over LA for the first time gave me an almost desperate feeling when I saw the smog over the metropolitan area. Our younger son was only one year old, and he soon caught asthmatic bronchitis during our visit. The advice from the paediatrician was clear and distinct; my son could only recover if we moved away from LA. Naturally this gave me a lot of guilt. To my relief, after we had moved back to Sweden after our year he recovered quickly. We had lived close to the Interstate 405 having four lanes in each direction and saw the demonstration of sources of the smog all the time. The gasoline was so cheap that it was not considered a real cost and was regarded almost as a birthright. The gas mileage of the cars was lousy, typically 12–13 miles/gallon (or 20 litres per 100 km). That was acceptable because gas was cheap and abundant. We paid 28 cents/gallon. In Sweden the price was around 2.5 times higher at the time. Taking the price index into consideration today's price is around 5.5–6 times more expensive than the LA price in 1970 and the Swedish gasoline price is still some 2.3–2.5 times higher than the US price. Also today, transportation is a major source of greenhouse gases in the US economy. The spirit of the times in the USA may be illustrated by a speech by President Richard Nixon in 1973: *'We use 30 percent of all energy ..... That isn't bad; that is good. That means that we are the richest, strongest people in the world and that we have the highest standard of living in the world. This is why we need so much energy, and may it always be that way.'*[7]

In 1970 we learned about the city of Seveso in Italy, some 20 km north of Milan. The name became a synonym for highly toxic dioxin, caused by another case of careless industrial production. At least 2000 people in a large area of the Lombard Plain were made sick by a dioxin cloud after an accident at a chemical plant in Seveso. Some 80 000 animals were slaughtered to keep the poison from entering the food chain. More than 10 years later Europe adopted the Seveso Directive, regulating the manufacture and storage of hazardous materials.[8]

In 1973 I began my research on water and wastewater treatment automation. Design and operation of wastewater treatment plants had traditionally been carried out only from a steady state point of view. In other words, only average loads and flow rates were considered, not the significant dynamic variations of the load over hours, days, and weeks. My control approach was considered with great suspicion by 'traditional' civil engineers, claiming that I did not really understand the real problems. Learning about water problems got me much more interested and involved in environmental thinking.

We had been aware, from World War II and during the Suez crisis in 1956, of the geopolitical consequences of oil. In the fall of 1973, we experienced another oil crisis. This time the OAPEC countries (Organization of Arab Petroleum Exporting Countries) proclaimed an oil embargo. The embargo was a consequence of the October war in 1973 between Israel and a coalition of Arab states, mainly Egypt and Syria. The Sinai Peninsula and the Golan Heights had been occupied by Israel in 1967. Egypt had the initial aim to gain control of the eastern bank of the Suez Canal to get full control of the Canal. On 6 October, the Jewish holy day of Yom Kippur (this is the Day of Atonement, a most important holiday in the Jewish faith), the Arab forces launched a surprise attack against Israel, both towards Sinai and the Golan Heights. The OAPEC embargo was aimed at nations that supported Israel during the October war. The embargo lasted until March 1974, causing the oil price to increase almost three-fold. This caused an oil crisis with several consequences on global economy and politics.

The 1973 oil crisis has been followed by more oil shocks, having impacts all over the world. The Iranian revolution in 1979 caused just a 4% drop in oil production. Still the political uncertainty caused the oil price to almost double over the coming year. In 1980 Iraq's oil production dropped significantly because of the Iran–Iraq war. This also created economic recessions all over the world. However, the development of solar photovoltaic and wind power over the last few years raises the hope that energy will become less influenced by geopolitical forces, when the 'fuel' is available all over the place. This is discussed elsewhere.[9]

In 1973 Ernst F. Schumacher (1911–1977) published his landmark book *Small is Beautiful*.[10] Schumacher was an economist from Oxford and described our economic systems and their purpose. He challenged the current state of excessive consumption in our society. He wrote and talked about 'natural capital' and chose fossil fuels as one example. We are treating them as income although they are undeniably capital items. He explained:

If we treated them as capital items, we should be concerned with conservation: we should do everything in our power to try and minimise their current rate of use; we might be saying, for instance, that the money obtained from the realisation of these assets – these irreplaceable assets – must be placed in a special fund to be devoted exclusively to the evolution of production methods and patterns of living which to not depend on fossil fuels at all or depend on them only to a very slight extent. … We are not in the least concerned with conservation; we are maximising, instead of minimising, the current rates of use; and, far from being interested in studying the possibilities of alternative methods of production and patterns of living – so as to get off the collision course on which we are moving with ever-increasing speed – we happily talk about unlimited progress along the beaten track, of 'education for leisure' in the rich countries, and of 'the transfer of technology' to the poor countries.'

We had the oil crisis in 1973 in vivid memory.

Too many economists had made us believe that all problems can be solved by printing more money. Yet, money cannot be eaten. At the end, someone must grow the food that we are eating. We burn fossil fuels and use resources that we have not produced. It is like spending money that we won in a lottery. Schumacher warned that the modern world must balance economic growth with the human costs of globalization. In 1995 *Small is Beautiful* was ranked as one of the 100 most influential books published since World War II'.[11] Schumacher's ideas are still my source of inspiration.

In 1974 I got acquainted with another important message: *The Limits to Growth*.[12] A group of 30 individuals from ten countries had gathered in Rome, at the instigation of Dr Aurelio Peccei, an Italian industrial manager affiliated with Fiat and Olivetti,[13] to discuss the present and future predicament of human beings. The participants were scientists, industrialists, educators, economists, humanists, and civil servants. Out of this meeting grew The Club of Rome.

They initiated their study under the leadership of Professor Dennis Meadows at Massachusetts Institute of Technology (MIT). They got a lot of inspiration from Professor Jay Forrester (1918–2016) of MIT, a pioneer in systems engineering. In 1971 Forrester had presented his book *World Dynamics*, including an extremely simple mathematical dynamical model of the world, consisting of just five differential equations.[14] His previous publications included books on *Industrial Dynamics* and *Principles of Systems*, describing the foundations of systems engineering. In *World Dynamics*, world development was described by the five variables: (1) world population, (2) pollution, (3) amount of irrecoverable natural resources (such as fossil fuels), (4) world capital investment, and (5) percentage of capital invested in the agricultural sector.

Forrester was already famous for his research in electrical and computer engineering during the 1940s and early 1950s. Forrester had invented the addressing system for digital computer memory. In the early 1950s he was appointed as head of Project Whirlwind, the team at MIT that built one of the first digital computers. Forrester figured out a way to organize the magnetic cores that stored information into a grid so that their contents could be retrieved. This was a true foundation of the modern computer.

Forrester was almost surely aware of the limitations of his world model. He used birth and death rates for high-income and low-income countries, based on available statistical data. Obviously, he did not predict all the advancement made later in medicine and public health. He calculated the impact of pollution but could not know how water treatment would progress. Many of the parameters and coefficients of the model are truly time varying, giving the predictions great uncertainty. Now, 50 years later we know several parameters that Forrester could not have known. For example, he predicted that the global population would peak at a little more than 5 billion in 2020. However, Forrester did not pretend to present quantitative predictions of the fate of Mother Earth. Rather, he wished to demonstrate the power of system dynamics to make predictions. The world would certainly collapse as a result of unlimited growth.

Dennis Meadows and his students developed a slightly more complex world model,[15] but the general performance was like Forrester's model. The book *Limits to Growth* was published in 1972 and received enormous attention. However, I was sceptical of the predictions – having not included uncertainty – to the year 2100 based on data from the first half of the 20th century. The qualitative message was clear but by showing quantitative simulations as if they were true reduced the credibility of their message. Notably, Forrester himself did not participate in this study. It taught me how dangerous it is to present quantitative results that are not well founded. Trust in the rest of the message may be lost, even if it was obvious that we cannot have unlimited growth, just as Schumacher and Forrester had said.

In 1975 I spent a sabbatical semester at the University of Houston, Texas. Professor John F. Andrews (1930–2011) was my host in Houston and one of the true pioneers in understanding wastewater treatment processes as dynamic systems. This was a most rewarding visit where I learnt a lot of process technology from John while he eagerly wanted to learn more about my experiences of automatic control. This led to a life-long cooperation and friendship. I had brought a picture from a Swedish weekly magazine to my colleagues in Houston, showing our capital Stockholm. In the foreground a number of people were swimming in Lake Mälaren in the centre of Stockholm, where the Town Hall was seen in the background. The clean water in downtown Stockholm drew a lot of attention and was a clear demonstration that biological wastewater treatment around Stockholm had been so successful that the lake was sufficiently clean for swimming. This picture became a main justification for partly locating the next International Water Association (IWA) specialized conference on instrumentation, control, and automation in Stockholm in 1977. Forty years later we would summarize the development and the experiences in instrumentation, control, and automation in wastewater treatment.[16]

During my stay in Houston a special symposium was organized on the theme 'limits to growth' with only invited people. I had the fortune to represent the University of Houston. Dennis Meadows and his colleagues in the Club of Rome were invited speakers to the symposium. Their serious message disturbed me not only by their confident predictions but by the fact that the team arrived in large limousines to talk about limits to growth. I learnt the lesson that *we must live the message.*

Also E. F. Schumacher was attending the symposium, and for me he made a long-lasting impression. His message was that we must build our economies around the need of communities and people, not of corporations. This was quite a provocative message in Houston, the oil and gas metropolis that was expanding like never before. Since his presentation in Houston in 1975 the CO_2 emission rate has increased about 2.5 times.

From Schumacher, I also got my first insight into sustainability, even if I did not use the term in 1975. He mentioned a second aspect of our limited resources: the tolerance margins of nature. Since my encounter with Schumacher, I have studied many aspects of our limited planet: water resources and water scarcity, polluted waters, dams from mining operations, limited resources of vital substances like phosphorus, and waters and wetlands that have been devastated by oil operations. They are all aspects of the sustainability concept. Several of them are included in earlier books and publications on water and energy.[17]

Rachel Carson and Paul Ehrlich had made us aware of the interconnections between the environment, economy, and social well-being. In 1972 the UN Conference on the Human Environment took place in Stockholm and naturally got special attention in Sweden. The conference was partly motivated by the pollution and acid rain problems of northern Europe, and it led to the establishment of the UN Environment Programme (UNEP). Naturally the oil crisis in 1973 fuelled the limits to growth debate. It was not until 1987 that the World Commission on Environment and Development (WCED) (called the Brundtland commission) formulated the concept *sustainable development* to general acceptance.[18]

Despite the world oil crisis in 1973, energy was still cheap in Texas in 1975. Air conditioning was truly necessary in the summer heat with 37°C and 100% humidity. During a weekend trip we went to a pizza place to eat. We experienced the temperature shock of entering the cool restaurant. The illumination was dim and after a while I discovered a fireplace with a real fire burning! The idea was to make the room a little cosier. The combination of air conditioning and fire became my symbol of energy waste. Very few people around me in Houston at that time understood my excitement about the waste of energy.

Over the years there has been a multitude of definitions of sustainability and there is still no scientific or political agreement on how to define sustainable development. My attitude has been that there is a combination of necessary developments to preserve our earth, our environment, and the resources that we have to use. It is closely related to our lifestyle and our consumption. For many people 'sustainable development' is defined like 'democracy' in the sense that it is 'universally desired, diversely understood, extremely difficult to achieve, and won't go away'.[19]

How does our economic system impact the way we live? Does it really affect what we truly care about? Schumacher seriously questioned the idea of unlimited growth, 'more and more until everybody is saturated with wealth'. We know that the environment will not be able to cope with the burden of growth. Modern economy is driven by greed and envy and Schumacher claimed that the inevitable result is 'nothing else than a collapse of intelligence'.

Exxon, the large Oil and Gas Company, was aware of climate change as early as 1977, 11 years before James Hansen told the US Congress about climate change in 1988 (see below). This sobering message was delivered by senior scientist James Black.[20] Still, this did not prevent Exxon from refusing to acknowledge climate change for decades and even worse, promoting climate misinformation. They had learnt from the way the tobacco industry had spread lies regarding risks of smoking.[21] They knew, once the world understood the risks, their products would not remain profitable. In 1978 Black warned Exxon that a doubling of CO_2 in the atmosphere would increase the global average temperature by 2–3°C. This is in good agreement with scientific results today. His warning came more than 40 years ago: *'present thinking holds that man has a time window of 5–10 years before the need for hard decisions regarding changes in energy strategies might become critical.'* Not many people listened. Exxon supported the creation of the Global Climate Coalition in 1989 (disbanded in 2002) to question the scientific basis for concern about climate change. It also helped to prevent the USA from signing the Kyoto Protocol in 1998 to control greenhouse gases. Actually, the USA with Al Gore as Vice President did sign the Protocol, but Congress did not ratify it. The USA had put as a condition for their signature the introduction of marked-based mechanisms (emission trades), while Europe was for planned control with regulations and investments. The Kyoto Protocol was criticized as having several inconsistencies and outright design flaws that seriously hinder its effectiveness. Later, in 2009, James Hansen officially criticized the Kyoto Protocol as an inefficient and indulgent 'cap and trade' system. Detailed criticism had been published earlier by Gianguido Piani.[22] Hansen stated in 2009 that 'the developed nations want to continue basically business as usual, so they are expected to purchase indulgences to give some small amount of money to developing countries'. Hansen preferred a 'carbon tax', not the Kyoto 'cap and trade' system.

Also, hydropower comes with an environmental cost. My father was a building contractor, so I was introduced early to building houses. Even if he did not build roads and bridges, they caught my interest very early, so as a 6-year-old first-grader I had heard about the Royal Institute of Technology in Stockholm and I made up my mind to 'become an engineer and earn a lot of money'. Early on I admired large hydropower dams. The marvellous engineering work to build these huge structures to provide electrical power for us made a great impression on me. In my country, Sweden, some 45% of our electrical power is today produced by the hydropower systems in the north of the country and then transmitted via huge power transmission lines to the main consumers in the south. I still remember when my father in the 1950s showed me the power line coming from the Harsprånget plant in the far north providing power for us more than 1000 km further south. At that time the power line was the world's longest power line with 380kV (now 400 kV) voltage. It was many years later that I realized that hydropower also comes with an environmental cost. In 1975 the Banqiao Dam (in Henan Province, China) failed due to extraordinarily heavy precipitation caused by a typhoon, combined with poor quality construction of the dam, which was built in the early 1950s. The flood immediately killed over 100 000 people, and another 150 000 died of subsequent epidemic diseases and famine.[23]

1980 global population: 4.5 billion.
Global CO_2 concentration: $\cong 280 + 59 = 339$ ppm

My first real challenge to discuss science with the general public took place in 1980. The nuclear power issue had become a true watershed in Swedish politics. The Harrisburg Three Mile Island nuclear accident had taken place in the year before and caused an increasing opposition against nuclear power. The Swedish Government arranged a referendum on nuclear power. Having worked as a nuclear engineer I got invited to take part in some public discussions. I did not want to argue for or against nuclear power, but instead wanted – to the best of my knowledge – give scientific reasons both for and against, inspired by the physicist Richard Feynman.[24] My first disappointment was that so few experts wanted to get involved in the debates. Some of them were prevented by their employers. Many engineers did not wish to discuss the issue since they claimed that they did not understand the complexity of nuclear power: 'I just work with nuclear core material', or 'I only know about turbine issues', etc. So, most discussions took place between laymen. I also realized that the scientific method is a handicap in a debate before an audience that is not an expert in science. The audience did not like 'on one hand' and 'on another hand'. They wanted a clear-cut message. So I realized that a lawyer-style message will win over a scientist-style message. This became a life-long lesson: as professional researchers we must learn how to communicate science results to the public and policy makers.[25] At the time I believed that knowledge is power and should decide the logical and ethical grounds on which we want to be making our decisions, but it seems more correct to say that money is power. This was expressed years later by James Hansen (see below).

The industrial disaster in Bhopal (state capital of Madhya Pradesh, India) in 1984 was another reminder how greed caused insufficient considerations for people's safety and environmental health. It was called an *'organized irresponsibility'* and took place in a Union Carbide factory making pesticides. The disaster was the logical consequence of poor plant design, lack of maintenance, and very low consideration of staff safety together with economic stress. Water had leaked into a storage tank containing 43 tons of methyl isocyanate (MIC), which started a reaction. This reaction is exothermic – the reaction rate increases with temperature – so the reaction rose quickly, and the mixture started boiling. The gas cloud, containing cyanide and the colourless poisonous gas phosgene, having a great inhalational toxicity, spread across the sleeping city, injuring more than half a million people. Estimates of the death toll range from 3700 to 16 000 people. Not only was the air deadly polluted, the groundwater outside the plant area, where large slum areas had developed, had become toxic. As a result of the disaster there was an emerging environmental justice movement, and both local and national groups were formed to campaign for legislative reform. There was great disillusion with regulators' inability to protect the environment and public health.[26]

James Hansen, director of the NASA Goddard Institute for Space Studies and a world leading scientist on climate issues, had already worked with climate change reality in the 1980s. Speaking as a NASA scientist he issued a warning that the world is failing *'miserably'* to deal with the worsening dangers. He presented what's considered the first warning to a mass audience about global warming in June 1988. At a US congressional hearing he declared *'with 99% confidence'* that a recent sharp rise in temperatures was a result of human activity. At the time I read about his warnings but did not understand the extent of the issue. More than 20 years later he expressed his frustration with greenwashing: *'politicians are happy if scientists provide information and then go away and shut up'.*[27] He also described how his testimony was misunderstood and misinterpreted. In his book Hansen expresses his experiences of climate debates:

> *'The scientific method, in one sense, is a handicap in a debate before a non-scientist audience. It works great for advancing knowledge, but to the public it can seem wishy-washy and confounding: 'on one hand this, on the other hand, that.' The difference between scientist-style and lawyer-style tends to favour the contrarian in a discussion before an audience that is not expert in the science.'*

Hansen has continued his brave battle.

On one of the last days of April 1986 my wife and I had a picnic and collected small nettles to make nettle soup. Returning home we listened to the news that the Swedish nuclear plant Forsmark had an emergency shut down because of leaking radiation. Days later it was discovered that the true source of radiation was far away, the Chornobyl nuclear reactor. Prevailing winds brought the nuclear fallout from Ukraine to Sweden and infected vegetables, such as nettles, with caesium 137. We ate our soup before we became aware of the radiation damage.

Nuclear energy accidents are rated from 1 to 7 on the International Nuclear Event Scale. There are only two nuclear disasters rated 7, Chornobyl and the 2011 Fukushima Daiichi nuclear disaster in Japan. Naturally, the nuclear accidents have influenced not only the opinion about nuclear energy but also investments in new plants. Nuclear energy has been presented as an energy source without a carbon footprint, but as the climate crisis has become apparent for the general public, the resistance to nuclear power has also increased.[28] By 2022, however, there are great movements to make nuclear power attractive again.

In 1987 I got a full professorship in Industrial Automation at Lund University and the research at the Department of Industrial Electrical Engineering and Automation forced me – and inspired me – to address automation problems in water systems as well as in electrical power systems. I started to appreciate that there are significant couplings between the disciplines, but it took me several years to get a deeper understanding of the interactions. In a way, this reflects the academic research problem. In order to attract financial support, one has to be highly specialized and use the right buzzwords. Furthermore, to explore couplings between research areas is a significant effort, and therefore one must work with the problems for a long time before a solid basis for the research can be formed.

In 1988 the UN Environment Programme (UNEP) and the World Meteorological Organization (WMO) established IPCC, the Intergovernmental Panel on Climate Change. This was endorsed by the UN General Assembly in the same year. The first IPCC Assessment Report was published in 1990 and formed the background information for the UN Framework Convention on Climate Change (UNFCCC) and the Rio conference in 1992.

By 1989 the CO_2 concentration in the atmosphere had reached 350 ppm. In November 1989, Margaret Thatcher, for the last time in her official capacity as the British Prime Minister, presented a speech to the UN General Assembly on environment and climate change.[29] Considering that Mrs Thatcher had a background in chemical research, she had a deep understanding of the problems we faced and presented ways to resolve them: *'The result is that change in future is likely to be more fundamental and more widespread than anything we have known hitherto. …. It is no good squabbling over who is responsible or who should pay. We shall only succeed in dealing with the problems through a vast international, co-operative effort.'* This was extraordinarily prophetic, and her words were even more shocking from the lips of a right-wing world leader who couldn't be dismissed as an irritable hippy. If the world had heeded her warning back then, imagine where we would be now? However, only weeks after her speech the Berlin wall fell, and the world got other priorities.

1990 global population: 5.3 billion.
Global CO_2 concentration: $\cong 280 + 74 = 354$ ppm

In June 1992 the UN Conference on Environment and Development (UNCED) took place in Rio de Janeiro, Brazil, and became known as the Earth Summit.[30] It was organized 20 years after the first Human Environment Conference in Stockholm, Sweden. High representatives from 179 countries were gathered in a massive effort to focus on the impact of human socio-economic activities on the environment. The interdependence of social, economic, and environmental factors was recognized, even if the term 'nexus' was invented only decades later. Success in one sector requires action in other sectors, so a systemic view is not a new invention. Integrating and balancing economic, social, and environmental concerns in meeting our needs is vital for sustaining life on Earth. This would require new ways of production, consumption, lifestyle, and decision making. The UNCED conference produced Agenda 21, an almost revolutionary programme how to invest in a sustainable future.[31] The principles formulated in the Rio Declaration are still valid. The Rio conference also decided to form the UN Framework Convention on Climate Change (UNFCCC). The report[32] states early that *'Acknowledging that change in the Earth's climate and its adverse effects are a common concern of humankind.'* Furthermore, the call for global cooperation is not new: *'Acknowledging that the global nature of climate change calls for the widest possible cooperation by all countries and their participation in an effective and appropriate international response, in accordance with their common but differentiated responsibilities and respective capabilities and their social and economic conditions.'*

Speaking to the British newspaper *The Guardian* in 2018, James Hansen said: *'All we've done is agree there's a problem. We agreed that in Rio in 1992 and re-agreed it again in Paris in 2015. We haven't acknowledged what is required to solve it. Promises like Paris don't mean much; it's wishful thinking. It's a hoax that governments have played on us since the 1990s'.*[33] Dr Hansen gave a lecture a few years ago and asked the audience: *'What would you do if you know what I know?'* Behind him on the screen he showed pictures when he was arrested by the police during a climate demonstration outside the White House. Can a researcher balance scientific integrity with political engagement? *Yes, I think it is necessary.* As the author Roger Pielke discusses in his book *The Honest Broker*[34] the researcher should have a role of 'honest broker', to clarify pros and cons of different actions instead of only present specific solutions.

A truly scientific debate should not be mixed up with political conflicts as I had experienced in the nuclear energy debate in the 1980s. Behind every diagram, table, or line of text, there are not only many hours of work. There is also a human being: we are parents, grandparents, friends, family members, or citizens. As a researcher I am also obliged to be a human. On this finite planet, we need to change the way we think, particularly about where and how we live, work, and travel, along with what and how much we consume. We have unmistakable indications that now is the time to make a change. Each one of us can play an important role in creating a world where we all live within our ecological limits. Even small actions can make a big difference.

James Hansen dedicated his book about the approaching climate crisis[35] to his grandchildren. A dominating driving force for me is my responsibility for my own grandchildren. *'Did you try to make a better world?'* Will we be considered good ancestors?

In the late 1990s my wife and I visited Morocco for a short holiday during the dark Swedish winter season. Instead of staying on the beach we wanted to discover more of the interesting country. We had heard about the Blue People, nomads from the Sahel region, so we went to Guelmim (Gulimin), found a local guide at the street, and continued to an oasis at the edge of the Sahara Desert. The Blue People had arrived on camels from far away and stayed in the oasis to trade goods they needed. A little pond of brown water in the oasis, around 10 metres across, provided the difference between life and death. We were invited into the tent of a proud representative of the people. This was one of the great memorable moments in my life. We had tea together and talked via our interpreter for hours. One of the first questions from our host was: 'Do you have water at home?' 'Yes', we replied. 'Do you have sufficient water for your cattle?' he continued. Thinking about the wealth of water in Sweden and our clean beautiful lake close to our little summer cottage I did not know how to answer the question properly, but said: 'Yes, the cattle have enough water'. I did not dare to mention that our lake has drinkable water. His immediate reply was: 'Why then did you come here?' His question has stayed with me since then: how would you properly answer his question and not feel guilty? Having clean water available is a sign of extravagant wealth. Too often we take it for granted.

Mining operations have a far higher cost than what is included in the price of the extracted mineral. Too often somebody else has to pay for disasters and for the clean-up of soil or polluted waters. All over the world there are abandoned mining sites causing severe environmental damage and human suffering.

In 1998 a mining accident took place at a mine located in Los Frailes, Andalusia, in southern Spain. The Swedish company Boliden was involved, since the mine was owned by Boliden's subsidiary, Boliden Apirsa S.L. A dam at Apirsa's mine breached, resulting in the release of 4–5 million m^3 of acidic metal-bearing water and tailings sand along the river Guadimar. The accident impacted large areas of agricultural land and affected areas 40 km downstream from the mining area. The clean-up operation took 3 years, at an estimated cost of €240 million.[36]

2000 global population $= 6.1$ billion.
Global CO_2 concentration: $\cong 280 + 89 = 369$ ppm

In the year 2000 we were made aware about still another accident – this time in Romania. The Baia Mare gold mine, operated by a joint Australian-Romanian venture, had spilled more than 130 000 m^3 of cyanide in four rivers, among them the Danube. It affected not only aquatic and plant life for long sections of the rivers but also drinking water sources for people in Serbia, Hungary, Romania, and Bulgaria for many months. Efforts from EU to ban the use of cyanide in mining have been blocked or ignored.[37]

In 2000 the UN Millennium Development Goals, time-bound and measurable goals for combating poverty, hunger, disease, illiteracy, environmental degradation, and discrimination against women, were presented. The goals were to be achieved by 2015. We realized that we had made a lot of improvements, but not enough, during the first 15 years of the new millennium.

I had worked with the water–energy nexus for a few years when the 17 UN Sustainable Development Goals (SDGs) were adopted by the international community in 2015 as part of the 2030 agenda for sustainable development. There are a lot of interlinkages between the various SDGs. Therefore, it is important to adopt an integrated approach towards their implementation, to serve not only as visions but also as guides. Sufficient energy and water will be needed to meet nearly all the development goals. SDG6 – clean water and sanitation – depends a lot on the availability of renewable energy, recognizing that much conventional energy generation today depends on the availability of water. SDG7 – access to affordable, reliable, sustainable, and modern energy for all – depends strongly on the development of renewables like solar photovoltaic and wind. Access to clean water and clean energy are closely coupled to the development of human health and well-being, environmental health and security.[38] How do we measure the success of the SDG?

During 2005–2010 I served as the editor-in-chief for three IWA journals, *Water Science and Technology*, *Water Supply*, and *Water Practice and Technology*. The contents of the journals reflect contributions from many of the 50 IWA specialist groups. It gave me the opportunity to get a basic appreciation of other specialist areas than my own research area of instrumentation, control and automation. This position helped me to further appreciate many couplings between various water aspects and energy, biodiversity, climate, and food.

Studying climate systems, I soon got to learn about three prominent Swedes who contributed significantly to the advancement of climate science: Svante Arrhenius, Bert Bolin, and Carl-Gustaf Rossby.

While studying physical chemistry in the 1960s, I did not realize Arrhenius' role in climate research, only that he had described how chemical reaction rates depend on temperature. In 1889 Arrhenius suggested a physical interpretation of this phenomenon which is now known as the Arrhenius equations. Arrhenius was professor in *physics* in Stockholm 1885–1905, and university rector (vice chancellor) 1897–1902. He became the first Swedish Nobel Prize winner when he was awarded the Nobel Prize in *chemistry* in 1903. Actually, Svante Arrhenius is considered the founder of physical chemistry together with Wilhelm Ostwald and Jacobus Henricus van't Hoff.

In 1896 Arrhenius published his ground-breaking paper 'On the influence of carbonic acid in the air upon the temperature on ground'.[39] He was the first scientist to model the greenhouse effect, that is how CO_2 will increase the heat content on earth. His calculations of the relationship between greenhouse gas concentration and temperature increase were amazingly close to the calculations made by the IPCC a century later. He estimated that if the CO_2 content in the atmosphere were to be doubled then the temperature would increase 5–6°C.

Professor Bert Bolin (1925–2007), a Swedish meteorologist, was another climate pioneer. He was one of the founders of IPCC and became its first chairperson in 1988–1998. Bolin was heavily involved in various international groups concerned with climate change in the 1960s. He was also the scientific director of the European Space Agency (ESA). He recognized that the emission of CO_2 was a key problem. Bolin organized several national and international conferences in Sweden in the 1970s and served as an advisor to Swedish governments and prime ministers, for example to Prime Minister Olof Palme in the 1970s and 1980s. Thanks to Bolin, Sweden became an important partner in climate work and Bolin's efforts have probably influenced Swedish energy policy. Bolin described many of his experiences in his own book.[40]

Carl-Gustaf Rossby (1898–1957) was a Swedish-American meteorologist and made significant scientific contributions in meteorology. He developed mathematical models describing large-scale air movement. In 1928 he was appointed professor at MIT, the first US professor in meteorology. He identified sinusoidal waves in the polar jet stream, now known as Rossby Waves. They can explain how global warming can cause extreme cold weather in the winter in the southern parts of the USA or in Europe. He worked on mathematical models for weather forecasting, the Rossby equations, forecasting the weather using an electronic computer in 1950. In 1950 he returned to Sweden and founded the Institute of Meteorology in connection with the University of Stockholm. He appeared on the cover of *Time* magazine in 1956.

I retired from my academic responsibilities in early 2006 and this opened new possibilities to think and study independently of financial support and peripheral academic duties. My primary goal was to further explore how water and energy are related. In my research I had worked in parallel with control and automation issues in water systems and in energy systems, but in separate silos. To some extent this is caused by the financing structures that have encouraged specialization rather than multidisciplinary cooperation. I realized that water professionals easily understood the importance of energy for all kinds of water operations. Energy people, however, most often took water availability for granted. Among my contacts, water scarcity was not an issue and therefore this was considered a non-issue.

Within a few years after 2006 there was an increasing interest in the water–energy nexus. The interest among water professionals had been there for some years, but many energy operators and energy professionals now considered the water–energy nexus crucial. A landmark report by Sandia Labs in 2006[41] inspired my own research.

In 2008 I visited Allan R. Hoffman in Washington DC. He is a solid-state scientist turning into a renewable energy and a water–energy professional, and served for many years as Senior Analyst, US Department of Energy, Washington DC. He had played a substantial role in the development of renewable energy while being a top scientific advisor in Washington. Allan had opened my eyes to the issue of water–energy nexus and inspired me in writing my first book on water and energy.[42] Since then, Allan and I have had regular contact and I have learnt a lot from him. In his book on *The US Government and Renewable*

Energy,[43] he describes how renewable energy was already an issue at the top political level in the late 1970s. Allan was instrumental in the effort to make President Carter decide to support renewable energy by adding a solar hot water heating system to the White House roof in 1979. The Reagan Administration had other priorities and would favour nuclear energy and fossil fuels. The solar panels were removed from the White House roof. The renewable energy R&D budget got reduced by a factor of eight from its height at the end of the Carter Administration.

A significant report on water and energy was published by the UN[44] in 2014. In the same year, water and energy was the special theme for the 2014 World Water Day in Stockholm and the US Department of Energy published a major report on the water–energy nexus.[45] I had the pleasure of being invited speaker both at the World Water Day in Stockholm and (later) the corresponding day in Oslo in 2014. The Stockholm World Water Week programme in 2014 was composed around the water and energy theme.

More than 15 years ago I met Oren Lyons, the Indian Chief from North America. We had invited him to Lund University. He made a lasting impression on me. Among other things he told me: *'If you do not have a moral question in your governing process, then you do not have a process that is going to survive.'* This is my deep motivation not only to understand the couplings between water, energy, food, lifestyle, and climate change but also to realize the impact of climate change, our living conditions, and basically the future of our children and grandchildren. Oren Lyons gave me new insights how our lifestyles influence the fate of our environment and our limited resources.

2010 global population $= 7.0$ billion
Global CO_2 concentration: $\cong 280 + 110 = 390$ ppm

Working in academic research with colleagues and students I have had the privilege to visit many countries around the world and live an 'everyday life'. In this way I have got some first-hand impressions of environmental problems. I have had problems breathing due to air pollution in Beijing, worried about water scarcity in northern China, watched how the long drought in Cape Town made the day zero real, felt the drought problems in Queensland, Australia followed by flooding, been confused how the monsoon in Malaysia was delayed, and with great frustration and anger learnt how inhabitants in the Niger Delta in Nigeria suffered from oil leakages and flaring. Still, being a visitor, I was not a victim.

The Deepwater Horizon oil spill in 2010 got huge attention, an event being closely monitored by powerful media. Around $780\,000\ \mathrm{m}^3$ of oil leaked out into the ocean, a spill that was three times the amount of oil leaking from the Amoco Cadiz disaster in 1978. The long-term oil leakages in the Niger Delta in Nigeria surpasses all these disasters having spilled out 2–3 times as much as the Deepwater Horizon. The Mexican Gulf and the Nigerian disasters have been described in detail elsewhere.[46]

The summer of 2018 was a wake-up call for many Swedes. In most of the southern half of Sweden it was the warmest summer on record. We experienced destructive forest fires all over the country. Climate change suddenly was no longer abstract: it became real and threatening.

If I had predicted in the summer of 2018 that in half a year, there would be a massive movement among young people to take climate change seriously and wake up the older generations, then only few would have believed me. And even more remarkable: this movement was inspired by a 15-year-old girl Greta Thunberg with pigtails, having Asperger syndrome, striking alone outside the Parliament building in Stockholm. Her demands were that the Swedish government reduce carbon emissions as stipulated in the Paris Agreement 2015. Unlike other climate demonstrators, Greta protested by sitting outside the Riksdag (Parliament) every day during school hours with the signboard *Skolstrejk för Klimatet* (school strike for the climate). Later in 2018 she would be speaking at the World Economic Forum, meeting the Pope, and meeting EU leaders. *Time* magazine named her as one of the world's 25 most influential teenagers of 2018. In 2019 *Time* magazine named her The Person of the Year. Today, in 2021 Greta has become an icon and role model for many young people that worry about the climate, notably at the COP26 meeting in Glasgow. Even Prime Minister Boris Johnson cited her in his introductory speech at COP26. *Her straight-forward message has consistently been: 'You have to listen to the scientists and take action.'*

2020 global population $= 7.8$ billion.
Global CO_2 concentration: $\cong 280 + 132 = 412$ ppm

We have a vivid memory of the August 2020 Beirut explosions. Ammonium nitrate, used predominantly in agriculture as a high-nitrogen fertilizer, was stored in the Port of Beirut. The massive explosion damaged buildings throughout the city and killed almost 200 people and injured more than 6500. The economic consequences for Lebanon are catastrophic. Such an accident could have been avoided without so much corruption and greed.

The UN General Assembly met on 21 September, 2021, and the Secretary General António Guterres presented a remarkable and razor-sharp speech to the world leaders.[47] From the complete talk I have chosen some sentences, illustrating his serious message:

'I am here to sound the alarm: The world must wake up. We are on the edge of an abyss – and moving in the wrong direction. Our world has never been more threatened. Or more divided. ..... The climate crisis is pummeling the planet. .... Human rights are under fire. ....Science is under assault. .... Solidarity is missing in action – just when we need it most. ... we are getting an F in Ethics. .... The recent report of the Intergovernmental Panel on Climate Change was a code red for humanity. We see the warning signs in every continent and region. ... We need a

45 per cent cut in emissions by 2030. Yet a recent UN report made clear that with present national climate commitments, emissions will go up by 16% by 2030. .... We are weeks away from the UN Climate Conference in Glasgow, but seemingly light years away from reaching our targets. We must get serious. And we must act fast. .... Like never before, core values are in the crosshairs. A breakdown in trust is leading to a breakdown in values. Promises, after all, are worthless if people do not see results in their daily lives. .... The problems we have created are problems we can solve.'

Surely, there are some positive technology developments. The efficiencies of a lot of equipment and industrial processes are improving all the time. Membrane technology has made tremendous improvement, influencing water treatment technology, including desalination. The costs of wind and solar photo-voltaic power are decreasing at an impressive speed, while the price for fossil fuel can only go upwards in the future, despite occasional price drops. Still, according to the International Energy Agency's (IEA) World Energy Outlook 2021[48] the rate of change to meet the climate crisis is far too inadequate.

doi: 10.2166/9781789062908_0001

Chapter 1

Introduction – setting the scene

During the last decades there has been an unprecedented progress in living conditions, public health, sanitation, child mortality, communication, and the global economy. But there is no such thing as a free lunch and progress comes at a cost. Still, our behaviour does not reflect such an understanding. We have treated the capital of the Earth as if it were free: coal, oil, gas, uranium, and minerals are extracted, and water is used wastefully, without concern that we are draining the capital. Agricultural land has been mishandled, and air, rivers, lakes, and oceans have too long been considered freely available and have served as bottomless waste bins.

1.1 WHERE WE ARE TODAY

A few numbers may illustrate how the conditions for the global population have developed during the last few decades. There has been impressive progress:

- Three quarters of the global population now have access to sanitation and to clean drinking water at home.
- World life expectancy has increased from 45 years in 1950 to almost 73 years in 2021.
- Around 80% of all children aged one year have been vaccinated for at least one disease.
- World literacy rate has increased from 67% in 1976 and to almost 87% today.
- The urban population has increased 4.3 times since 1960, much faster than the global population increase of 1.6 times. The urban population grew from 34% to 56% in the last 60 years.
- The number of deaths caused by natural disasters has *decreased* during the last century, not because of fewer extreme events, but due to better warning systems and better protection in many places.
- Four out of five people in the world have access to some electricity.

But still:

- More than 2 billion people have no access to clean water at home and more than 2 billion people lack access to basic sanitation, like toilets and latrines.
- More than 800 children younger than 5 years die every day from diarrhoea due to lack of clean water.
- More than 800 million people face chronic food deprivation.
- More than one billion people lack electrical power.
- Economic inequality is bizarre: in 2021 the world's richest 1% have more than double the wealth compared to 6.9 billion people or 88% of the global population.

1.2 HOW WE GOT HERE

Several developments have occurred, isolated or interconnected, over recent decades. Each one of them can give a picture of how we got here today[49]. Development and growth have come also with an environmental cost:

- Total CO_2 emissions increased from around 9.5 Gt/year in 1960 to almost 40 Gt/year according to the IPCC Sixth Assessment Report 2021[50].
- Electricity generation using fossil fuels has increased from less than 6000 TWh in 1960 to around 16 000 TWh in 2020.
- Carbon emissions are set to hit an all-time high by 2023. Just 2% of pandemic recovery finance was being spent on clean energy[51].
- Biodiversity loss is apparent today and is a clear and distinct way to measure the health of our planet. Wildlife is the canary not only of coal mines, but of the entire planet.
- Deforestation contributes to the increasing greenhouse gas concentration in the atmosphere.
- Land area for agriculture is around 38% of the global land surface. Global cropland area per capita was around 0.45 hectare in 1961 and has decreased continuously since then to 0.21 hectare in 2016[52]. There is a great risk that projected cropland expansion and intensification will take place in regions that are valuable for biodiversity conservation, such as tropical areas of Latin America, Central Africa, and South-East Asia.
- An average of 14 million people every year were reported displaced by extreme weather disasters in the period 2008–2014[53]. In 2021 increasing intensity and frequency of extreme weather events are already causing more than 20 million people to leave their homes[54].

Efficiency and productivity have been guiding stars of our industrial societies. We have looked at environment as something 'over there' and have not understood that we are part of it. Chief Oren Lyons expresses our relation to Earth like this: *'What you people call your natural resources our people call our relatives'*. We need to recognize that all things are connected. If Earth suffers then we will suffer. Therefore, it is imperative to comprehend the water–climate–energy–food–economy–lifestyle nexus. Let us look at some of the developments:

- *Coal* was the remarkable resource that made the industrial revolution possible. Then we detected the cost for coal mining: mine accidents, water pollution, mine explosions and fires. We suffered from air pollution caused by coal burning and became aware of its climate impact. We paid for the coal but did not pay for the damage that it caused.
- *Oil* was the 'clean' replacement for dirty coal and so much easier to handle. It was the basis for the internal combustion engine and made our mobility possible. However, it also caused accidents during exploration, leakages during transportation, water pollution, destroyed wetlands and marine life, threatened the subsistence of farmers and fishermen, and led to geopolitical conflicts and wars. Again, the price of oil does not take the costs for accidents and environmental damage into consideration.
- *Natural gas* became the cleaner alternative to coal. So convenient to transport in pipelines. However, we had to recognize that natural gas flaring and methane leaking from pipelines and oil wells have a huge impact on our climate.
- *Nuclear power* promised to satisfy our electrical energy thirst for many generations. Then Three Mile Island, Chornobyl, and Fukushima reminded us about its vulnerability. Moreover, we did not find a dead certain solution to the final disposal of uranium.
- *Food production* was made more efficient by using irrigation, fertilizers, and machinery at large scale, but it led to water scarcity. Monocultures and too much irrigation impoverished the soil. As hand pumps were replaced by electric pumps for irrigation, groundwater levels shrank to alarming levels.
- *Food habits* in the rich part of the world have changed rapidly. Meat production has increased four-fold since the 1960s. This is problematic since meat is an 'inefficient' food source. Food waste, both in the low-income and in the high-income world, but for different reasons, is unacceptable. The food on our breakfast or dinner table has often travelled long distances from other parts of the world.
- *Industrial production* has been the sign of increasing welfare. However, waste products and harmful components leaked into the environment, polluted the air, caused damaged water sources, and poisoned fish, which in turn led to unsafe food and human suffering.
- *Mobility* is greater than ever. We gained freedom never seen before using our cars and aircraft. Now the price has become apparent in terms of air pollution, greenhouse gas emissions, and lack of resources.
- *Economic inequality* is greater than ever, both within nations and between countries. The idea that we can grow indefinitely in a finite world is absurd. Only a minority of the global population can enjoy all the benefits of our progress. We designed economic systems that favoured rich people and affluent nations.

- *Impatience* – in the high-income world we expect instant reward:
    - *Electrical energy*: even on the coldest winter day or the hottest summer day we take it for granted in wealthy countries to have access to all the electric power needed for machinery, heating, and cooling. This requires high power capacity, designed for the peak demand. The huge development in power electronics for electric drives has made pumps and compressors so much more efficient.
    - *Food*: we have got used to fresh fruit and vegetables being available all around the year. In my country Sweden we could buy apples from our antipode New Zealand at a lower price than local apples during our own harvest season. Cheap energy and transportation made it possible.
    - *Mobility*: in a country with four distinct seasons, we can afford to escape to warm destinations during unpleasant winter weeks. If the summer weather is not sufficiently pleasant, we may rush to get a more enjoyable summer elsewhere.

For too long we have perceived nature as something that needs to be conquered and be used to satisfy the needs of humans. We have come to a point in global development, where we are forced to recognize that everything is connected and that we are part of nature. If we are ever going to solve critical issues, we must appreciate the interconnections between our actions, consumption, economy, and nature. Systems thinking is one way to grasp all these connections.

1.3 GOING FROM HERE

To manage the complexity of the global resources we need to systematically analyse how they are coupled. In the rest of the book, we will try to describe where we are and why a systemic approach may help us. However, solutions depend not only on science and technology. We need to achieve a better understanding how we as humans set our priorities and what kind of moral compass that we have.

The goals or performance criteria of wicked problems like global resources, cannot be defined objectively. Instead, they are the result of difficult negotiations and compromises between different interests. However, as a control engineer, I look for some performance, so that I can compare achievements ('measurements') with the goal and find ways to reach corrective actions. Pledges being made at conferences and international meetings are nothing else than 'reference values' or 'desired values'. If they are not connected via 'measurements' to some action, they are meaningless and empty wishes.

1.4 SOME SCENARIOS

To describe the complex interactions between water, energy, food, economy, and climate we will try to follow some elements from cradle to grave, from source to final use and disposal. This will hopefully form a basis for a more systematic description of the couplings in the following chapters.

1.4.1 Fossil fuels

Every *minute*, around the clock, the world consumes:

- More than 10 000 m³ of oil
- Around 14 000 tons of coal, corresponding to 12 700–20 000 m³
- More than 7 million m³ of natural gas, corresponding to 5000–6300 tons (at 0°C and 1 atm or 100 kPa).

The volume of an Olympic swimming pool is 50 m × 25 m × 2 m = 2500 m³. Hence, if we filled the swimming pool with oil and supplied the world, it would be emptied in 14 seconds. Likewise, another pool filled with coal is emptied in 8–12 seconds. The famous Trafalgar Square in London has an area of 12 000 m². Natural gas produced (at normal pressure) will fill a volume to cover the whole square and rise to a height of 52 metres, to the top of the Admiral Nelson statue, in 5 seconds. Around the clock.

This is the culprit of global warming. Figure 1.1 is a schematic of the life cycle of oil, from extraction to final use.

Oil is extracted as 'conventional' oil (crude oil) or as 'unconventional' oil, usually extracted by hydraulic fracking (fracturing) or from tar sand. For the oil extraction, water is used, sometimes as steam to facilitate pumping, sometimes as heated water to get rid of bitumen in the tar oil. In hydraulic fracturing, water is pumped deep into the sediments at extremely high pressure to fracture the sediment where oil and gas are trapped. Several chemical components are added to the water, many of them toxic. Water is polluted (called 'used water' by the operators) and has an impact on the environment. Extraction produces CO_2 emissions from the energy-requiring operations as well as methane emissions caused by leaking natural gas. Flaring of natural gas contributes to both greenhouse gas emission and air pollution. There are apparent risks for large volumes of oil leaking out into the environment, on land or in water, due to accidents or to human errors.

The refining and processing of oil products will further require both water resources and energy. Transportation of oil, in pipelines or in ships, is not without risks. Pipelines may leak, and oil tanker accidents have taken place. Transportation requires energy, thus producing more emissions.

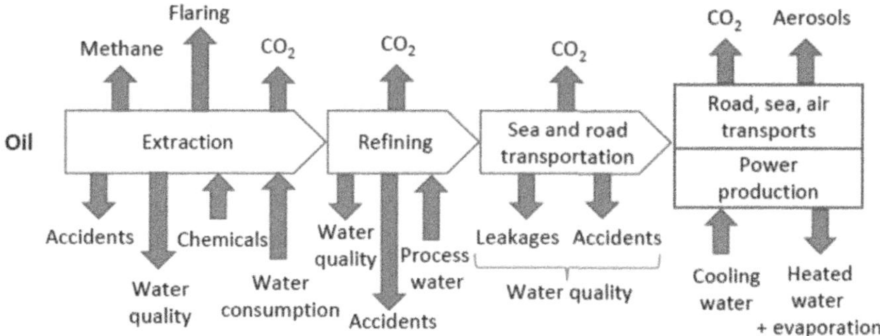

Figure 1.1 The life cycle of oil.

The final user of the oil product, burning the oil, be it for electric power production or for transportation, will cause further emissions of greenhouse gases as well as air pollution from particle emission. Cooling thermal power systems requires water that is primarily heated but also consumed due to evaporation.

Coal has a similar life cycle to oil. For centuries coal mining has been recognized as a dangerous operation, for workers and for the environment. Land is destroyed and water is contaminated. Coal is both the largest source of electricity generation and the largest single source of CO_2 emissions.

Natural gas has a large carbon footprint, since the main component is methane. Hydraulic fracturing is a vast source of natural gas production. Leakage of gas is a significant source of global warming risks. This happens both around gas extraction and in pipeline transportations. Natural gas is increasingly replacing coal as an energy source for power system operations.

1.4.2 Food

Every hour, around the clock, the world eats 40 000 tons of meat. For this consumption, more than 8 million animals are slaughtered. Some 95% of them are chickens, and then it includes 170 000 pigs and 35 000 cattle. Every hour.

Ruminant animals – cattle, sheep and goats – can eat grass, but are increasingly fed with cereals and soya to enhance the production rate. Figure 1.2 is a rough illustration of the life cycle of the meat that we eat.

Water is essential for the growth. A significant part of farmland is irrigated, and this is a huge threat to water resources in many places. Chemical fertilizers and pesticides are added, influencing both soil and water quality. Farming machinery contributes to the carbon footprint. Land use for the cattle raising is a significant cause of deforestation around the world. Cattle contribute to both water pollution and to methane emissions. Industrialized food processing has a significant footprint of both water and energy. Transportation and distribution look quite different in wealthy and in low-income countries. Significant amounts of food are transported long distances between continents and will, of course, contribute to food's large carbon footprint. In low-income countries, less availability of efficient transportation and safe storage will cause food waste. In wealthy countries, most of the food waste takes place close to the consumer.

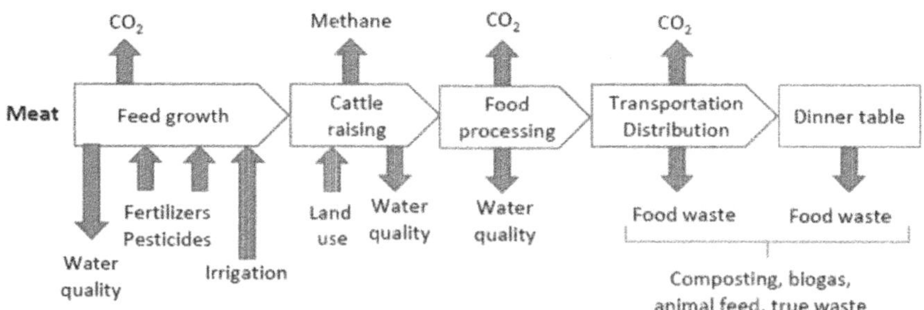

Figure 1.2 The life cycle of meat.

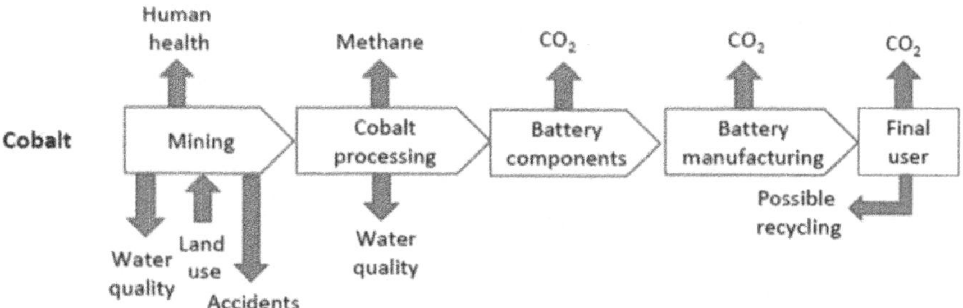

Figure 1.3 The processes from cobalt mine to car battery.

1.4.3 Renewable energy – critical minerals

Renewable energy comes with a lot of promises to replace fossil fuels. Being a variable source of energy, there will be an increasing requirement of storage capacity. Electric cars need efficient batteries. The need for critical minerals and metals like cobalt and lithium is apparent. Figure 1.3 indicates some of the interactions in the chain of processes from mining to battery manufacturing for cobalt.

Mining operations have a huge impact on water resources and water quality as well as on human health. The cobalt is processed in various steps and most of the cobalt mined today is transported from DR Congo to East Asia. Battery component and battery manufacturing have probably a large environmental impact, but much of this information is not openly available. Transportation causes a carbon footprint. The final user, the electric car owner, needs to charge the battery. If the electric power is coming from fossil fuels, then the electric car will cause CO_2 emissions.

1.4.4 Economics

Economic systems favouring unlimited growth in a limited world have been driving global warming. There have been serious warnings for at least three decades, but the influence on the economic system has been insignificant.

Economic inequalities in the world, between nations and between individuals, have created two groups. The wealthy part is responsible for the major contribution to global warming, while the less fortunate part has been subject to the most serious consequences. Economics and lifestyle are closely coupled to water use, energy use, food habits, and global warming.

1.5 OVERVIEW OF THE BOOK

To understand the impact of the global development it is necessary to consider not only each of them separately but to see how they are connected in a network, illustrated in the highly simplified diagram in Figure 1.4. Water scarcity problems cannot be solved in isolation but must be understood in relation to climate, energy, food, and lifestyle. Similarly, energy challenges must be solved taking the other 'components' into consideration.

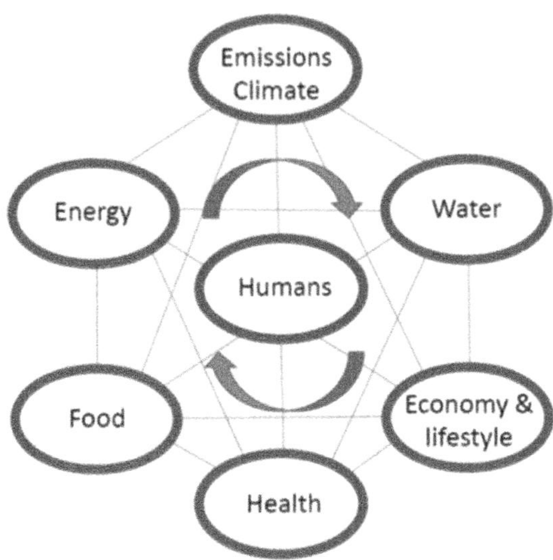

Figure 1.4 The complex problems that we face today cannot be solved in isolation. They are interconnected and influence each other bi-directionally.

	Climate	Water	Energy	Food	Health	Economics &Lifestyle
Climate	**Ch. 3–4**	4.3, 4.4, 4.6, 4.8, 5.1, 5.2	6.1, 6.3, 6.6, 6.10, 6.11	5.6, 7.1, 7.2	4.7, 4.12	7.3, 8.1, 8.2, 8.3, 9.2
Water	----	**Ch. 5**	5.2, 5.3, 5.4, 6.4, 6.5, 6.6, 6.7, 6.8, 6.9, 6.12, 6.13	5.6, 6.6, 7.3, 7.4	5.1, 5.7, 6.8	5.5, 5.7, 8.1, 9.4
Energy	----	----	**Ch. 6**	6.6, 7.3, 7.5	6.13, 6.14	8.1, 8.4, 9.4
Food	----	----	----	**Ch. 7**	7.6, 7.7	7.6, 8.1, 9.4
Health	----	----	----	----	3.7, 4.7, 4.12 5.5, 6.5, 6.8, 6.13, 6.14, 7.7	8.3 9.4
Economics & Lifestyle	----	----	----	----	----	**Ch. 8–9**

Figure 1.5 Graphical table of content of interactions. The matrix is symmetric, indicating that each interaction has a clear cause–effect direction. Here we have chosen to show them as bidirectional. For example, water has an impact on energy operations, and energy production has a consequence for water resources or water quality.

In the endeavour to present this complexity in some systematic manner each one of the couplings should be described while still not being able to demonstrate the complete picture. The reader will find out if this is accomplished in the book. Figure 1.5 is a graphical table of content, illustrating where the various interactions are investigated. For example, to find how water and energy are related we look at the connections both from a water and an energy perspective, and the interactions are analysed in sections 5.2, 5.3, 5.4, 6.4, etc. Similarly, the interactions between other components are indicated in the figure. Some sections, for example in Chapter 6 on energy or Chapter 8 on economics are descriptions that are specific for that topic, without apparent descriptions of interactions.

doi: 10.2166/9781789062908_0011

Chapter 2
Systems thinking

> *We find a shift in metaphors – a change from seeing the world as a machine to understanding it as a network.*
> Capra and Luigi Luisi (2019).

Complex problems cannot be solved by simple solutions. It is becoming more and more evident that major problems of today – global warming, climate change, water scarcity, clean energy availability for all, food for everybody, and economic inequality[55] – cannot be understood and solved in isolation. So, how do we approach this?

- Listen to and trust the scientists? A good idea, but it not sufficient. We also need to translate the messages in to actions.
- Make the politicians act right now? Great, but it does not solve the problem. We also need to act personally and do what we can do.
- Transforming our food systems to care for the planet? Yes, but far from sufficient. How can we from the wealthy part of the world support less fortunate populations to relieve their hunger?

Each one of the approaches is insufficient if it is done in isolation. Only together they are effective. In other words: *interdisciplinarity is using good general knowledge of many disciplines to solve issues that deep knowledge of one discipline cannot solve.*

The global problems that we face are systemic problems, which means that they are interconnected and interdependent. There are solutions to many of these complex problems, but they require a radical shift in our perceptions, our thinking, and our values. There is a formidable challenge: leaders and decision makers at all levels should see that many of the complexities are interrelated. We must 'connect the dots'. Our solutions today will influence future generations.

Systems thinking= everything is connected

Climate change and all its consequences are truly complex. Various aspects of it are studied by different disciplines. Consequently, they come to isolated conclusions. When each academic discipline, action group, industry, or political leader talks about 'climate change' they could be referring to drastically different things. A marine ecologist worries about increasing ocean temperature, while a power generation engineer talks about increasing hydropower limitation. They address the same issue. However, if they do not listen to each other, they think that their problem is the dominating challenge. By listening to perspectives from different disciplines our minds can be truly extended. It also means that to solve extremely complex problems, we must rely on and trust others. This includes not only so-called experts, but also *everyday people*. The latter often know more than many scientists about consequences of climate change and scarcity of water, energy, or food.

In Figure 1.4 we illustrated that each of the components climate, water, energy, food, economy, and lifestyle has to be understood from a systems perspective, but the complex system must be approached with detailed knowledge of the components. Systems thinking involves connectedness, relationships, and context. The global problems must be understood from the interactions and relationships between the parts. We have to develop many more cross-sectoral interactions, from academic curricula to research, development, financing, and politics. The essence of systems thinking is captured in the phrase, 'the whole is more than the sum of its parts'. The idea is not new: the origin of this phrase is to be found in Aristotle's *Metaphysics*, more than 2,300 years ago.

2.1 SYSTEMIC APPROACH

Integration is the core of systems thinking. The same underlying idea of 'systems' can denote technical, social, economic, and living systems. The meaning of systems derives from the Greek *syn+histanai*, meaning 'to place together'. To understand things systemically literally means to put them together to establish the nature of their relationships.

A fundamental principle of systems thinking is that everything is interconnected. Humans need food, air, and water to sustain our bodies. Trees need CO_2 and sunlight to thrive. Everything needs something else, often a complex array of other things, to survive. The systems thinking perspective defines a fundamental principle of life. We need to change our way of seeing the world as consisting of a huge number of system components to viewing it as a dynamic, interconnected array of relationships and feedback loops. The limits to growth discussions in the 1970s, as described in the prologue, had the systemic view that everything is connected. Even if the detailed calculations were only partially correct, the approach showed the importance of interconnections. It also reminds that systems thinking must be combined with deep knowledge of the components.

Essential to understand relations and patterns

Systems thinking deals a lot with relationships and patterns, and how various components interact with each other. Therefore, we cannot deal with environment, technology, people and their priorities and ambitions, as independent issues. All of them are interconnected. Sometimes we need a component view and other times we need a systems perspective, like looking at objects from the ground or from a helicopter. The true challenge is to understand the component and process aspects from a systems perspective and to comprehend the system aspects from a component perspective.

> *Analysis* means taking something apart to understand it; *systems thinking* means putting it into the context of a larger whole.

Systems thinking is the opposite of analytical thinking. Analysis means *taking something apart* to understand it; systems thinking means *putting the parts into the context of a larger whole.* A dominating methodology in science has been analysis, where we dissect complex problems into manageable components. This reflects the mechanical and reductionist view, where the complexity is broken down into parts. Synthesis aims to combine two or more things to create something new. It is about understanding the whole and the parts at the same time, along with the relationships and the connections that make up the dynamics of the whole. Systems thinking got a lot of attention from Jay W. Forrester in the early 1970s[56], as we discussed in the Prologue.

There is abundant evidence that the mechanical/reductionist paradigms have had an adverse impact on our ability to address all categories of our challenges. Machine thinking has mindlessly destroyed so much; we need a different mind-set to repair the damage. The degradation of environment has carried us to the boundaries of what Earth can take. We will not be able to fix the damage to the environment using the same paradigm. As we diligently extract non-renewable resources, we are also greatly diminishing and endangering the system that created life on this planet.

> We are part of the ecosystem

Systems thinking and a living network paradigm enable us to see ourselves, our communities, and our environment as components of a larger ecosystem, and to recognize the importance of interactions between different systems. Some examples may illustrate the connections:

- Turning production to low-income countries means that they may be blamed for higher emissions that are primarily caused by consumption in higher income countries. This was a key criticism of the Kyoto Protocol.
- Decreasing over-consumption in rich countries will naturally influence production. This in turn depends on material and energy resources.
- Increasing renewable energy production should mean that less fossil fuels would be needed. However, cutting off fossil fuel burning too abruptly will cause transitional problems. This relates to critical minerals or metals, land use for feedstocks for biofuels, and competition with growth of food resources and water.
- Decreasing meat consumption is good both for the climate and for water resources. However, this change will not only affect the cattle industry but also the farmlands producing feed for the cattle. Ultimately, it would mean that the land area required for a certain number of calories will change.

- Public health problems are closely related to the access to clean water and is also related to water storage for energy or irrigation.
- Decreasing urban water consumption means that water age (the time that water remains in the distribution system) will increase. This may influence the ultimate water quality.
- Increasing industrialization in a low-income country will increase the competition for water.

> Looking at climate change risks, multiple drivers interact, as do the risks themselves.

We live in a highly networked world. Looking at climate change risks, multiple drivers interact, as do the risks themselves. It is obvious that climate change has wide implications, even if it may be characterized primarily by the CO_2 content and the major emission source, fossil fuels. Climatologists emphasize that real-world experience underscores the complexity of interactions among multiple drivers of climate change risk[57]. Not only the risks have to be assessed, but the interactions that generate the risks must be identified. Potential impacts due to climate change as well as responses to climate change relate to risks. To approach complex systems like these it is crucial to think across sectoral and regional boundaries, and risks are not only related to physical phenomena but also to socio-economic drivers.

> Lifestyle has a profound impact on global warming

Climate change has impacts on water resources. Lack of water resources has a great influence on food production. Energy has consequences not only for climate but depends on available water, for example, for cooling thermal plants, for hydropower, and for finding metals for battery production. The priorities of financing institutions have a great impact on climate change or on alternative energy sources. The lifestyle of wealthy people has great significance for energy and water consumption and, consequently, on climate. A measure of success in the western world is mostly related to income and consumption. If happiness is believed to be correlated with capital wealth, it will be too hard to change lifestyle.

Connections between technological, environmental, and socio-economic systems require that we understand how risks are transmitted from one system or sector to another. This in turn creates new risks or amplifies existing ones. Most often it is convenient to break analysis into silos, taking a component-oriented, rather than interaction-oriented, view. This is to some extent done also by the Intergovernmental Panel on Climate Change (IPCC) which divides its assessment into three separate working groups focused on: (1) physical climate change; (2) climate impacts, vulnerability, and adaptation responses; and (3) emissions mitigation. This approach is useful for synthesizing thousands of discipline-specific studies and reflects the largely sectoral approach of

> *'All things are bound together, all things connect. Whatever befalls the earth, befalls also the children of the earth'.*
> Oren Lyons

many governments. However, by considering individual sectors, assessments can miss important interactions[58].

2.2 FEEDBACK

Feedback is an essential component of systems thinking. There are several feedback mechanisms related to global warming and climate, such as how global warming causes methane release due to the permafrost thaw. This will in turn amplify the global warming. Such feedback is an example of positive feedback, leading to the growth of its effect and instability. In a drought situation more groundwater may be used for irrigation. This may led to energy waste, which in turn affects climate. Furthermore, unrestricted use of groundwater will lower the water table, increasing irrigation challenges, demanding even more electrical power. Food habits with more meat consumption often leads to more land use, increasing water consumption and other negative impacts. Meeting extreme temperatures with more air conditioning (as suggested by a policy maker) will further amplify climate change.

Receiving data and comparing them with desired outcomes is the basis for feedback and corrections and is an essential part of our decision making and learning. As the few examples above illustrate, water problems or the energy challenges cannot be solved in isolation. Norbert Wiener (1894–1964), the true pioneer of cybernetics (from Greek *kybernetes*, 'steersman') defined cybernetics as the science of 'control and communication in the animal and the machine'. Early on in his work on cybernetics, Wiener was aware that feedback is an important concept for modelling not only living organisms but also social systems. Norbert Wiener's milestone publication from 1948 defines the basic principles of cybernetics that has later developed into control and communication[59]. He wrote: *'It is certainly true, and the social system is an organization like the individual, that is bound together by a system of communication, and that it has a dynamics in which circular processes of a feedback nature play in important role.'*

The principles of feedback control are applicable all the way up to high-level strategic decisions. The framework is always the same, whereas the measurements, the analyses, and the decisions are different. Understanding feedback loops is intimately connected to the concept of causality. One event or dynamical change results in another change in a dynamic and constantly evolving system. Cause and effect are fundamental concepts in physical systems and in life in general. To understand how components in a system influence each other we must practice causality concepts and systems thinking.

> Principles of feedback control are applicable all the way up to high-level strategic decisions

2.3 WICKED PROBLEMS

At planning, strategic, and policy levels, there are numerous complex challenges that have been characterized as 'wicked' problems[60] This is a social, cultural, or organizational problem that is difficult or impossible to solve. Knowledge about it may be incomplete or contradictory. There are many people and opinions involved, or the economy is a major obstacle. Often the problems are interconnected with other problems. Therefore, there is no apparent correct solution.

We may replace the term wicked with VUCA (volatile, uncertain, complex, ambiguous) problems to address the complex but solvable problems more correctly with a new mindset. To handle VUCA problems requires wisdom, which is a process of continuous learning, a journey rather than a destination.

> Handling wicked problem requires wisdom, a process of continuous learning,

Conventional problems often have scientific solutions developed by experts. A wicked problem definition depends on stakeholders' views and perspectives. For conventional problems, we can often judge if the outcome is true or false, or if it is successful or unsuccessful. For a wicked problem there is mostly no 'correct' solution. The outcome may be better, worse, or acceptable.

> Acquiring wisdom is an incremental process of integrating different types of knowledge, opinions, and interests into a holistic framework.

Dealing with complexity requires wisdom: and acquiring wisdom is an incremental process of integrating different types of knowledge, opinions, and interests into a holistic framework rather than mastering the details of a single aspect of knowledge. Achieving wisdom is a process of learning: a journey rather than the destination. To handle many of the problems described in this book requires that people from many different disciplines, cultures, social status, and economic conditions should cooperate. This is probably the challenge of our time.

2.4 HYPEROBJECTS

Since the nuclear bombs over Hiroshima and Nagasaki, we live in the nuclear age, the time when humans can destroy all life on Earth. We who grew up in the 1940s and 1950s can still remember the arms race, the fear of a nuclear winter and radioactive radiation. At any time, a senseless leader could push the button. This was another kind of fear compared to climate change. The climate threat has a low intensity and has been on-going for many years but is getting less attention. It is difficult to relate to a specific place and to one responsible leader. It is global and appears in many ways. The nuclear bomb threat was highly explosive and could be released at a specific moment.

The two global threats are seldom connected, but neverthless, both are real. The Anthropocene age is changing the Earth. The concept of *hyperobjects* provides one potentially useful way of seeing, comprehending, and adapting to these grand problems. The idea of hyperobjects has been conceived by Timothy Morton, Professor of Contemporary Philosophy at Rice University,

Houston, Texas[61]. Global warming is the hyperobject par excellence. Global warming is not a traditional object, because it cannot be seen directly but only through data about it. And even so it cannot be perceived as one whole object. As Morton expresses it: *'The panic and denial and right-wing absurdity about global warming are understandable. Hyperobjects pose numerous threats to individualism, nationalism, anti-intellectualism, racism, speciesism, anthropocentrism, you name it. Possibly even capitalism itself.'*

> Global warming is not a traditional object. It cannot be seen directly but only through data. It is a hyperobject.

Morton explains that the problem with hyperobjects is that you cannot experience one, not completely. You also can't *not* experience one. They bump into you, or you bump into them; they bug you, but they are also so massive and complex that you can never fully comprehend what's bugging you. This oscillation between experiencing and not experiencing cannot be resolved. It's just the way hyperobjects are.

In the 1950s the fear of nuclear war was raised also via movies. Today, sea level rise, loss of biodiversity, or carbon footprint are not understood in the same way. Our language is too meagre to describe what is happening around us. At the same time, it is remarkable that we believe that we will never experience a nuclear war, even though there are around 14 000 nuclear warheads ready to be fired. In July 1946, as the USA planned to test the first hydrogen bomb at the Marshall Islands the American military leaders explained for the local residents: *'We are testing these bombs for the good of mankind and to end all wars.'*

Do we see the correlation between the attitude to the Marshall Islands citizens and our disrespect for the poorest in the world that are now affected by our lifestyle and consumption?

2.5 THE UN SUSTAINABLE DEVELOPMENT GOALS

The 17 UN Sustainable Development Goals (SDGs) were adopted in 2015 by the international community as part of the 2030 agenda for sustainable development[62]. It should be recognized that there are a lot of interlinkages between the various SDGs. Therefore, it is important to adopt an integrated approach towards their implementation.

Nearly all the development goals depend on sufficient energy and water being available. There are strong couplings between water and energy. SDG 6 – 'Ensure availability and sustainable management of water and sanitation for all' – relies on available and affordable energy. This links to SDG7 – 'Ensure access to affordable and reliable, sustainable, and modern energy for all'. Conventional energy like thermal power plants and hydropower depend heavily on water availability.

> Without clean water and affordable energy none of the SDGs can be fulfilled.

This reliance can be significantly decreased using solar photo-voltaic and wind power system. The strong dependency between SGD6 and SDG7 are increasingly recognized[63]. The role of renewable energy and water solutions will directly and indirectly contribute to all the other 15 SDGs. This has been elaborated in a previous book by the author[64].

> Making renewable energy and water solutions available will directly and indirectly contribute to all the other 15 UN Sustainable Development Goals.

2.6 GLOBAL RISKS

The Global Risks Report 2020, published by the World Economic Forum[65], presents a detailed and sobering account of global risks, estimated for the next 10 years. The risk assessment is based on an annual Global Risks Perception Survey, completed by approximately 800 members of the Forum's diverse communities (called the *Multistakeholders*). For the first time the report also presents a survey among around 200 people, representing the younger generation of emerging global social entrepreneurs and leaders (the *Global Shapers Community*).

As illustrated in Table 2.1, all the top five global risks for *both* categories, in terms of *likelihood*, are in the environmental category. In the *impact* group three out of the five are categorized as environmental. Not surprisingly, the Forum's younger constituents show even more concern, ranking environmental issues at the top in both the short and long term, and they see the existential risks not only to their generation but to the wider global community. The water crises are naturally connected to environmental as well as society and conflict risks.

Table 2.1 Long-term risk outlook according to World Economic Forum (2020).

		Multistakeholders	**Global Shapers**
Likelihood	1	Extreme weather	Extreme weather
	2	Climate action failure	Biodiversity loss
	3	Natural disasters	Climate action failure
	4	Biodiversity loss	Natural disasters
	5	Human-made environmental disasters	Human-made environmental disasters
Impact	1	Climate action failure	Biodiversity loss
	2	Weapons of mass destruction	Climate action failure
	3	Biodiversity loss	Water crises
	4	Extreme weather	Human-made environmental disasters
	5	Water crises	Extreme weather

doi: 10.2166/9781789062908_0019

Chapter 3
Climate today

> '*I am here to sound the alarm: The world must wake up. We are moving in the wrong direction. Our world has never been more threatened. Or more divided.*
>
> *The climate crisis is pummeling the planet…. Economic lifelines for the most vulnerable are coming too little and too late – if they come at all…. The climate alarm bells are also ringing at fever pitch.*
>
> *We see the warning signs in every continent and region. Scorching temperatures. Shocking biodiversity loss. Polluted air, water, and natural spaces. And climate-related disasters at every turn.*
>
> *When they see billionaires joyriding to space while millions go hungry on earth…*
>
> *When parents see a future for their children that looks even bleaker than the struggles of today…*
>
> *And when young people see no future at all…*
>
> *Solidarity is missing in action – just when we need it most. We passed the science test. But we are getting an F in Ethics…..We must get serious. And we must act fast. The problems we have created are problems we can solve.*'
>
> Secretary General António Guterres
> UN General Assembly, 21 September 2021

There is an amazingly simple characterization of climate change, identified by a single number, the CO_2 concentration, measured in parts per million (ppm). Similarly, any progress in climate actions can be monitored in terms of CO_2 concentration. It is remarkable that a small concentration of about 400 parts per million has such a significant influence on the global temperature. Only 4 molecules out of 10 000 air molecules are CO_2. If there were no greenhouse gases in the atmosphere the global temperature would be around $-15°C$. Because of the greenhouse effect created by CO_2 and water vapour, the average temperature of the Earth is around $15°C$, which allows for life to exist.

Water vapour is also a greenhouse gas, the most abundant one. It absorbs longwave radiation and radiates it back to the surface, thus contributing to warming. Unlike other greenhouse gases, however, water vapour stays in the atmosphere a much shorter time, typically for days (before precipitating out), while CO_2, methane, and other gases will stay from years to centuries. Furthermore, the addition of water vapour to the atmosphere is typically not caused by human activities but is a feedback process. Warmer air can hold more moisture. Thus, the increase in water vapour contributes to even more warming, so the water vapour enhances the greenhouse effect.

> *To get a feeling for the low CO_2 concentration, imagine that we could collect all CO_2 to atmospheric pressure in one layer around the globe. This layer would be only around 4 metres thick.*
>
> *Does it sound believable that humans can influence such a thin layer by our emissions?*

Gases that undergo chemical reactions, like ozone or ozone-forming chemicals like nitrous oxides are relatively short-lived. CO_2, however, stays in the atmosphere for a long time. Conventionally it is assumed that CO_2 remains in the atmosphere for 100 years. What our generation adds will influence many human lives in the future.

In July 2014 NASA launched the OCO-2 (Orbiting Carbon Observatory) satellite. It gathers global CO_2 data with high precision and resolution and observes the whole Earth. From space it makes around 100 000 measurements of atmospheric CO_2 every day. In 2019 the OCO-3 was launched, further providing crucial atmospheric data. OCO-2 is in a polar orbit and OCO-3 is mounted to the International Space Station, circling Earth from 52° north to 52° south latitudes. This is where most of Earth's living things are found. OCO-3 can collect a denser dataset than OCO-2 over high-carbon regions such as the Amazon rainforest.

The UNFCCC parties (United Nations Framework Convention on Climate Change) have arranged annual formal meetings to assess progress in dealing with climate change. The first COP (Conference of the Parties) meeting was held in Berlin in 1995 and COP26 was recently completed in Glasgow in 2021.

In the 1990s the conferences negotiated the Kyoto Protocol (decided in 1997) and from 2011 the COP meetings negotiated the Paris Agreement in 2015. The negotiations are extremely complex since any final text of a COP must be agreed by consensus.

Two words had never been mentioned in the final documents of the COP meetings, until COP26 in 2021: *fossil fuels* (is it a coincidence?). The cause–effect can be expressed in this dramatically simple relationship. Still the consequences are extremely complex.

The increased concentrations of CO_2 in the atmosphere and in the oceans are a result of accumulated emissions since the industrial revolution. This is sometimes overlooked in the negotiations where various nations blame each other for the current crisis. CO_2 emissions from fossil fuel combustion were practically zero before 1750. The UK was the first industrialized nation and the first fossil fuel CO_2 emitter. In 1751 its emissions, as well as those at global level, were less than 10 million tons (Mt). Today they are 3600 times higher.[66] Figure 3.1 shows the accumulated emissions from the ten countries with the highest impact.

> The words 'fossil fuel' have not been mentioned in COP documents.

The seriousness of the issue among climate scientists was expressed by Raymond Pierrehumbert, lead author of the 2018 IPCC report: *'Let's get this on the table right away, without mincing words. With regards to the climate crisis, yes, it's time to panic.... We are in deep trouble.'*[67]

3.1 THE 1992 RIO CONFERENCE

The scientific community has been aware of climate change and its dramatic consequences for decades. The Keeling curve had developed into a clear warning sign of climate change since 1958. Prime Minister Margaret Thatcher had warned the UN General Assembly in 1989 and in a resolution 44/228, adopted by the General Assembly on 22 December 1989,[68] it was decided to

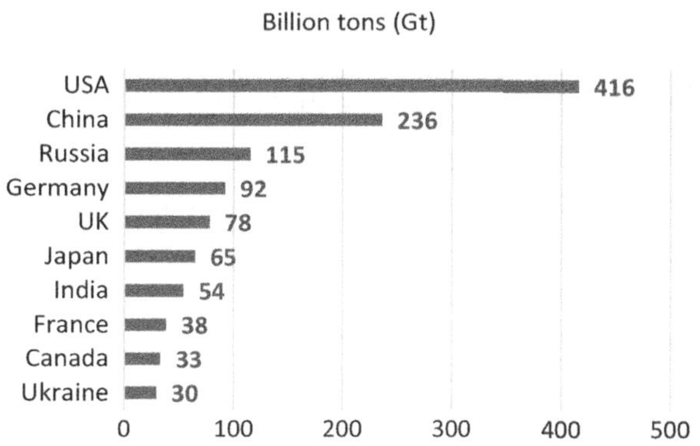

Figure 3.1 Cumulative emissions of CO_2 between 1750 and 2020.

convene the UN Conference on Environment and Development, to coincide with World Environment Day, in June 1992. *'The Conference should elaborate strategies and measures to halt and reverse the effects of environmental degradation; … to examine the state of the environment and changes that have occurred since the UN Conference on the Human Environment, held in 1972, … to recommend measures to be taken at the national and international levels to protect and enhance the environment, … to examine the relationship between environmental degradation and the international economic environment.'*

The principle of international responsibility concerning climate change, documented in the *UN Framework Convention on Climate Change,*[69] decided in the Rio conference in 1992 is crystal clear:

Principle 3.1: *'The Parties should protect the climate system for the benefit of present and future generations of humankind, on the basis of equity and in accordance with their common but differentiated responsibilities and respective capabilities. Accordingly, the developed country Parties should take the lead in combating climate change and the adverse effects thereof.'*

> *The Parties should **protect the climate system** for the benefit of present and future generations of humankind.*
>
> UN Rio 1992

Principle 3.2: *'The specific needs and special circumstances of developing country Parties, especially those that are particularly vulnerable to the adverse effects of climate change, and of those Parties, especially developing country Parties, that would have to bear a disproportionate or abnormal burden under the Convention, should be given full consideration.'*

Principle 3.3: *'The Parties should take precautionary measures to anticipate, prevent or minimize the causes of climate change and mitigate its adverse effects. Where there are threats of serious or irreversible damage, lack of full scientific certainty should not be used as a reason for postponing such measures, taking into account that policies and measures to deal with climate change should be cost-effective so as to **ensure global benefits** at the lowest possible cost.'*

> *Where there are threats of serious or irreversible damage, lack of full scientific certainty should not be used as a reason for postponing such measures*

Among the commitments, these may be emphasized:

Article 4.4: *'The developed country Parties and other developed Parties included in Annex II shall also assist the developing country Parties that are particularly vulnerable to the adverse effects of climate change in meeting costs of adaptation to those adverse effects.'*

Article 4.8: *'In the implementation of the commitments in this Article, the Parties shall give full consideration to what actions are necessary under the Convention, including actions related to funding, insurance and the transfer of technology, to meet the 15 specific needs and concerns of developing country Parties arising from the adverse effects of climate change and/or the impact of the implementation of response measures, especially on:*

a *Small island countries;*
b *Countries with low-lying coastal areas;*
c *Countries with arid and semi-arid areas, forested areas and areas liable to forest decay;*
d *Countries with areas prone to natural disasters;*
e *Countries with areas liable to drought and desertification;*
f *Countries with areas of high urban atmospheric pollution;*
g *Countries with areas with fragile ecosystems, including mountainous ecosystems;*
h *Countries whose economies are highly dependent on income generated from the production, processing and export, and/or on consumption of fossil fuels and associated energy-intensive products; and*
i *Land-locked and transit countries.*

….. The Parties shall take full account of the specific needs and special situations of the least developed countries in their actions with regard to funding and transfer of technology.'

Yet, in the three decades since Rio the world has emitted more CO_2 than during the entire century before. And many of the catastrophes James Hansen warned about in 1988 have become a reality. As Fredi Otto, climate scientist at the University of Oxford, expressed it recently: *'All of this is happening exactly as we have known it would happen.'*

> Many of the catastrophe warnings from 1988 have been a reality. *'All of this is happening exactly as we have known it would happen.'*

3.2 CLIMATE AS REPORTED BY IPCC IN 2021

IPCC, the Intergovernmental Panel on Climate Change, was established in 1988 and the first assessment report was published in 1990. The mission of IPCC is *'to provide policymakers with regular scientific assessments – every six or seven years – on the current state of knowledge about climate change.'* The IPCC does not carry out original research but rather serves as a clearinghouse for assessing and synthesizing the relevant literature. Thousands of scientists contribute to writing and reviewing the IPCC reports, which are then reviewed by governments. The IPCC is an association of volunteer scientists who produce arduous technical assessments, not policy recommendations. IPCC is scrupulous in recognizing any uncertainty in all its estimates. Its targets for the needed level of emissions reductions are never presented as a single figure but in terms of ranges and probabilities.

The sixth assessment report (AR6), published in 2021, formed a basis for the COP26 negotiations. Despite the concerns by the IPCC authors, the scientific evidence from AR6 was explicitly acknowledged in the Glasgow Climate Pact. The authors had reasons for their concern: the IPCC special report on 1.5°C warming was published before the 2018 COP24 in Katowice, Poland. The report was just 'noted' in the final document and no conclusions from the report were acknowledged.

In my summary of the IPCC findings, I have excluded the ranges and the probabilities and refer to the main AR6 report for all details. In short, we have more than enough information to take decisive actions based on what we know, fully aware of uncertainties presented.

The Physical Science Base report in 2021[70] of almost 4000 pages is an overwhelming and convincing documentation of the current climate situation. The report is a high-level summary of scientific results concerning climate, the role of human influence, and information about certain regions of the world. The AR6 cites more than 14 000 scientific papers and IPCC has received around 78 000 comments. It should be emphasized that all the results in AR6 are approved by all the authors. There are further scientific findings in the literature, but if there is no unanimous agreement of them, they are not included in the AR6.

The report states that *'It is unequivocal that human influence has warmed the atmosphere, ocean and land. Widespread and rapid changes in the atmosphere, ocean, cryosphere and biosphere have occurred.'* Note that the greenhouse gas (GHG) concentration measurements are well-mixed and can represent the global situation. In 2019, atmospheric CO_2 concentrations were higher than at any time in at least 2 million years, and concentrations of methane and nitrous oxide were higher than at any time in at least 800 000 years. Since 1750, increases in CO_2 (49%), methane (162%), and in nitrous oxide (23%) concentrations far exceed the natural multi-millennial changes between glacial and interglacial periods over at least the past 800 000 years. As of August 2021, the GHG concentrations had reached:

> IPCC: 'it is unequivocal that human influence has warmed the atmosphere, ocean and land'.

- CO_2 – 410 parts per million (ppm)
- Methane (CH_4) – 1866 parts per billion (ppb)
- Nitrous oxide (N_2O) – 332 ppb.

IPCC is now clear: human influence on the climate system is an established fact. The report claims that *'human influence has warmed the climate at a rate that is unprecedented in at least the last 2000 years'*. IPCC has further concluded (with various but high likelihoods) that human activities have been the main drivers of several signs of climate change:

- *Global surface temperature*: each of the last four decades has been successively warmer than any decade that preceded it since 1850. Global surface temperature was 1.09 [0.95 to 1.20]°C higher in 2011–2020 than during 1850–1900. Land warming, ice loss, and atmospheric warming accounted only about 5%, 3%, and 1%, respectively of the heating, while ocean warming accounted for 91% of the heating in the climate system. The land surface will continue to warm more than the ocean surface and the Arctic will continue to warm up more than global surface temperature, above two times the rate of global warming. Natural land and ocean carbon sinks are projected to become less effective, that is, the proportion of emissions taken up by land and ocean will decrease with increasing cumulative CO_2 emissions. As a result, a higher proportion of emitted CO_2 will remain in the atmosphere. Compared to preindustrial times, over the past 5 years 8–11% of the globe had exceeded 2°C warming and in Sweden the year 2020 was 3.3°C warmer.
- *Precipitation:* changes have been observed since the mid-20th century.
- *Snow cover*: there has been a decrease in the Northern Hemisphere spring snow cover since 1950.
- *Ice sheets*: there has been a global retreat of glaciers since the 1990s, a decrease in the Arctic Sea ice area, and the observed surface melting of the Greenland Ice Sheet over the past two decades. In 2011–2020, annual average Arctic Sea ice area reached its lowest level since at least 1850. Late summer Arctic Sea ice area was smaller than at any time in at least the past 1000 years. The global nature of glacier retreat since the 1950s, with almost all the world's glaciers retreating synchronously, is unprecedented in at least the last 2000 years. Permafrost thawing, loss of seasonal snow cover, of land ice, and of Arctic Sea ice will be amplified by the additional warming. The Arctic is likely to be practically sea ice-free in September at least once before 2050.
- *Ocean warming*: The global upper ocean (0–700 metres) has warmed since the 1970s. The world ocean, in 2021, was the hottest ever recorded by humans[71] and the last 5 years have had the warmest seas since records began. There is a year-to-year variation of the ocean heat content primarily tied to the El Niño-Southern Oscillation. Despite the La Niña oscillation in 2021 that cooled the Pacific Ocean, the record temperature was reached. Changes in ocean circulation play important roles locally, but the predominant changes result from human-related changes in atmospheric composition. Warmer oceans supercharge weather systems, creating more powerful storms and hurricanes. Warmer oceans lead to a warmer and moister atmosphere, promoting more intense rainfall in all storms. This will increase the risk of flooding. Warming oceans also threaten marine ecosystems, like coral reefs and fisheries.
- *Ocean salinity:* The low pH value as observed in recent decades, is unusual in the last 2 million years. There is a pattern of observed changes in near-surface ocean salinity.
- *Sea level rise*: the global mean sea level increased by 0.20 metres between 1901 and 2018. Global mean sea level has risen faster since 1900 than

over any preceding century in at least the last 3000 years. Thermal expansion explains 50% of sea level rise since the 1970s, while ice loss from glaciers contributed 22%, ice sheets 20% and changes in land-water storage 8%. The rate of ice-sheet loss increased by a factor of 4 between the 1990s and the 2010s. Icesheet and glacier mass loss were the dominant contributors to global mean sea level rise during 2006–2018. In the longer term, sea level is committed to rise for centuries to millennia due to continuing deep-ocean warming and ice-sheet melt and will remain elevated for thousands of years. Over the next 2000 years, global mean sea level will rise by about 2–3 metres if warming is limited to 1.5°C, 2–6 metres if limited to 2°C.

> Are we planning our civilisation to last for millennia?

After the IPCC report, further evidence of ocean temperature increases have been reported.[72] Monthly sea-surface temperatures from 150 years, from 1870 through 2019, were mapped and the location and time of extremes were recorded decade by decade. By comparing monthly instead of annual averages more in detail it was detected how the oceans are warming, and an increasing number of extremes were detected. From 2014 on, more than half the surface water areas of the oceans are warmer than the most extreme events from 1870 to 1919. Extremes of yesterday are now the new normal.

3.3 CLIMATE PIONEERS

We may recall the results published by Svante Arrhenius in 1896. Using the Stefan Boltzmann law, he formulated his original greenhouse law, expressed mathematically by:

$$\Delta T = \alpha \cdot \ln(c \, / \, c_0)$$

where ΔT is the temperature change, c the CO_2 concentration and α and c_0 constants. The formula is still useful today. Arrhenius estimated that a doubling of CO_2 would cause a temperature rise of 5–6°C. Later he adjusted the value downwards to 1.6°C (including water vapour feedback: 2.1°C). Estimates from IPCC, a hundred years later say this value (the climate sensitivity) is likely to be between 2 and 4.5°C. Arrhenius expected CO_2 levels to rise at a rate given by emissions in his time, so a CO_2 doubling would take about 3000 years; since Arrhenius time the increase has been almost 50%.

It is also appropriate to pay attention to the 2021 Nobel Prize winner in physics, Syukuro Manabe. He pioneered climate models to be simulated in computers in the 1960s. Considering the computing power at that time he was forced to simplify to make the models possible to compute. Only the key processes were included, such as solar radiation, water vapour, CO_2, and a reflecting land area. Still he was able to predict the temperature increase to 2–3°C when the CO_2 concentration in the atmosphere had been doubled, a similar range to Arrhenius's as well as IPCC's models.

For me as a Swede it is particularly interesting to note that the scientific background material published with the 2021 Nobel Prize mentions Svante Arrhenius 14 times, just 4 fewer than Manabe, who actually won.

Quite naturally the current climate models are enormously more complex and include more greenhouse gases, vegetation, sea currents, geological processes, and human activities. Also, all the CO_2 that we emit into the atmosphere does not stay there but is partly absorbed by vegetation and the oceans. Still the largest supercomputers do not have sufficient capacity to simulate the climate with the same degree of details as the weather forecasting models.

The other 2021 Nobel Prize winner Klaus Hasselmann has together with Syukuro Manabe built a basic understanding of climate modelling. Hasselmann wrote a model in the 1970s that helped scientists understand both how weather and climate interact and how to diagnose humanity's role in heating the atmosphere. Hasselmann has warned us about global warming for 50 years. Now he is pinning his hope on the young generation and trusts that the decision makers will listen to them. Professor Manabe testified in the US Congress in 1988 together with James Hansen. Manabe, however, talked with a heavy Japanese accent and now believes that the politicians did not understand anything that he said. Still James Hansen was 99% sure that human activities had already caused global warming. Hansen's testimony caused a lot of media attention. Nevertheless, hardly anything happened.

The third Nobel Prize winner Giorgio Parisi has also contributed to understanding climate change, even though he had another viewpoint, explaining complex physical systems. His results in complex theory have made it possible to understand and describe many different and apparently entirely random material and phenomena (such as weather) and relate this to a better understanding of long-term phenomena such as climate change. Knowledge about the climate rests on a solid scientific foundation.

The fact that decision makers now consider climate research more seriously is probably thanks to IPCC, since the first report in 1990. In 2007 IPCC and Al Gore shared the Nobel Peace Prize.

Actually, global warming was discussed in a popular TV show in Sweden in 1969 called 'Ask Lund' where researchers from my own Lund University were answering questions from a panel. The CEO of Volvo, Gunnar Engellau, asked if increasing concentrations of gases from human activities would influence the global temperature. My former colleague Bertram Broberg (1925–2005) gave a most insightful answer how both water vapour and CO_2 will influence the global temperature.

3.4 THE MILLION-YEAR PERSPECTIVE

Temperature and CO_2 variations during the last 800 000 years have been carefully recorded, using measurements from ice cores from Greenland and Antarctica. There is a clear relationship between temperature variations and CO_2 concentrations. The rising and falling CO_2 levels coincide with the onset of ice ages (low CO_2) and interglacial periods (high CO_2). There is a temperature and CO_2 peak around every 100 000 years, the natural changes between glacial and interglacial periods. These periodic fluctuations are caused by changes in the Earth's orbit around the sun, the precessional movements of the Earth axis, and the position of the moon. This link between global temperatures and greenhouse gas concentrations – especially CO_2 – has been valid throughout Earth's history. There is a time lag between atmospheric concentration variations and temperature changes. Consequently, even if we finally do manage to stabilize atmospheric concentrations, temperatures will continue to slowly rise for years or decades.

Over the period of 800 000 years, the CO_2 concentration varied mostly between around 180 and 280 ppm and for most of the period atmospheric concentrations of CO_2 did not exceed 300 ppm. From the industrial revolution to the end of the 1980s, the CO_2 concentration had risen by around 70 ppm (keep in mind the warning by Margaret Thatcher in 1989). For the last 30+ years we have seen an additional rise of around 70 ppm and the atmospheric concentrations is now approaching 420 ppm. Looking at the long-term concentration diagram, it seems that the CO_2 and temperature peaks were reached quickly. However, these 'rapid' increases usually took place over around ten millennia. Now we have seen the CO_2 concentration increase even more in a couple of generations, at least a factor 100 faster time scale. This is the core of the problem: nature – which we are part of – has too little time for adaptation.

3.5 THE CO_2 BUDGET

From a physical point of view the answer is straight-forward. It is obvious that it is necessary to limit cumulative CO_2 emissions in combination with other greenhouse gases, in particular methane. IPCC has confirmed that there is near-linear relationship between cumulative anthropogenic CO_2 emissions and the increase in global surface temperature.

Over the period 1850–2019, a total of 2390 ± 240 (likely range) $GtCO_2$ of anthropogenic CO_2 was emitted. Remaining carbon budgets have been estimated for several global temperature limits and various levels of probability, based on the estimated value of the so called TCRE (transient climate response to cumulative CO_2 emissions) and its uncertainty, estimates of historical warming, variations in projected warming from non-CO_2 emissions, climate system feedbacks such as emissions from thawing permafrost, and the global surface temperature change after global anthropogenic CO_2 emissions reach net zero.

To limit the temperature increase to 1.5°C, there is only 500 Gt left to emit from the beginning of 2020. Even so, there is only 50% probability that global warming will remain below 1.5°C (Table SPM.2, IPCC, 2021). If we would obtain 83% probability, then the budget shrinks to 300 Gt. With the current emission rate of 31.5 Gt/year this means that all CO_2 budget is used up before 2030. UNEP analysis[73] suggests the world is on course to warm around 2.7°C with hugely destructive impacts. In its report 'State of climate 2021' the World Meteorological Organization (WMO) states

> With current greenhouse gas emission rate all remaining CO_2 budget is used up before 2030.

that the build-up of warming gases in the atmosphere rose to record levels in 2020 despite the pandemic.[74] The amounts of CO_2, methane, and nitrous oxide rose by more than the annual average in the past decade. WMO finds that the national pledges to cut greenhouse gas emissions by 2030 are just 7.5% per year compared to the previous pledges made in 2015. To keep the 1.5°C goal alive would require 55% cuts by the 2030, a seven-times higher ambition.

Around half of emissions from human activity are absorbed by oceans, and to some extent trees and land. But this ability can vary significantly, depending on for example temperatures and rainfall. The Keeling curve clearly indicates the role of CO_2 absorption by vegetation: there is an annual oscillation of the CO_2 concentration with a maximum in the spring of the Northern Hemisphere. As summer approaches more CO_2 will be caught. NOAA (National Oceanic & Atmospheric Administration) results confirm that the amplitude of the annual oscillations is 4–5 ppm, corresponding to only 2 years of CO_2 base level increase. More vegetation is valuable for the climate, but it does not solve the climate crisis. Even more important, it is not correct to compare just carbon masses and not to consider their chemical bindings. To compensate for example burned aircraft fuel with trees in the Amazonas is an oversimplification. Carbon in the fuel has high-energy bindings unlike the carbon in the plant mass.

WMO confirms that the CO_2 level in 2020 was 149% of the pre-industrial level. The last time the Earth experienced a comparable concentration of CO_2 was 3–5 million years ago, when temperatures were 2–3°C warmer and sea level was 10–20 metres higher than it is today.

The National Academies of Sciences, Engineering and Medicine in the USA have assessed the latest in climate science, technology options, and socioeconomic dimensions and provide advice how to reach the goal of net zero emissions by 2050.[75]

3.6 THE COP26 AGREEMENTS

Current policies to reduce, or at least slow down, growth in CO_2 and other greenhouse gas emissions will have some impact on reducing future warming. But if the aim is to limiting warming to 'well below 2°C' – as is laid out in the Paris Agreement – the world is clearly far off-track. In Glasgow the countries of the world were called upon to improve their pledges in 2022 to satisfy what had been decided in the Paris agreements in 2015. The G20 countries have a particularly high responsibility, having a majority of the global economy. They are also responsible for some 75% of the global emissions.

The UN Secretary General Mr Guterres was talking a very clear language: *'Enough of brutalizing biodiversity, killing ourselves with carbon, treating nature like a toilet, burning, and drilling and mining our way deeper.'* He added *'we are digging our own graves'*, reminding us that our planet is changing before our eyes from melting glaciers to relentless extreme weather events. He also reiterated that sea-level rise is double the rate it was 30 years ago, that oceans are hotter than ever, and that parts of the Amazon Rainforest now emit more carbon than they absorb.

> *'Scientists are clear on the facts. Now leaders need to be just as clear in their actions.'*
>
> UN Secretary General António Guterres at COP26

The famous environmental activist and broadcaster David Attenborough said that if working apart we are a force powerful enough to destabilize our planet, working together, we are powerful enough to save it.

We may have the impression that climate conferences are organized so that the countries of the world meet to solve the climate crisis. This is only partially true. Countries are also there, and some countries are primarily there, to protect their national interests.

Some of the COP26 agreements and pledges are summarized here:

- *Deforestation*: leaders from more than 100 nations, representing about 85% of the world's forests, promised to stop deforestation by 2030. In a separate move, a group of high-income countries pledged US $12 billion for forest protection until 2025. There is no specific information how the funding will be provided.
- *Methane*: more than 100 countries agreed to cut 30% of current methane emissions by 2030. Methane is currently responsible for a third of human-generated warming. China, Russia, and India did not join, but there is hope that they will later. China and Russia are the methane top emitters with 18% and 10% respective of the total global methane emissions.
- *Coal*: more than 40 countries agreed to shift away from coal. This included major coal-users like Poland, Vietnam, and Chile. However large coal-dependent countries like Australia, India, China, and the USA did not sign. The agreement covered coal but no other fossil fuels like oil and gas.
- *Finance*: around 450 financial organizations agreed to back renewable energy and direct finance away from industries depending on fossil fuels. These organization control together some US$130tn ($10^{12}$). High-income countries were asked to double their collective commitments by 2025 compared to 2019, to address the climate crisis.
- *Climate fund*: the governments have failed to meet a 2009 pledge to provide US$100 billion per year in climate finance for low- and middle-income countries by 2020. It looks as if it will take until 2023 to reach this goal, and that around 70% of the finance will be provided as loans.
- *US and China*: the countries pledged to co-operate on climate over the next decade. This would include methane emissions, the transition to clean energy, and de-carbonization.

A group of high-income countries promised to end fossil fuel financing overseas. This effort could slow the growth of oil and gas projects in low-income countries. However, it would do nothing to decrease the development in top producing countries like the USA, Russia, Saudi Arabia, or Canada. Major funders of fossil fuel projects, like China, Japan, and South Korea, did not join the deal. Obviously, such a deal will further deepen the injustices of climate change. For example, low-income countries will not get the money to build gas powered plants while high-income countries are free to continue building them domestically.

For the most part, any commitments made at COP26 will have to be self-policed. Only a few countries are making their pledges legally binding. According to the Production Gap Report,[76] governments still plan to produce twice the amount of fossil fuels in 2030 than what would be in line with the global warming goal of 1.5°C. Aminath Shauna, the Maldives' Minister of Environment, Climate Change and Technology, expressed the fear that hard-hit nations have, facing an existential threat: *'The difference between 1.5 and 2°C is a death sentence for us. What is balanced and pragmatic to other parties will not help the Maldives adapt in time. It will be too late.'*

'The difference between 1.5 and 2°C is a death sentence for us'.

One month after COP26 the Glasgow Work Programme was published.[77] This is a 10-year plan, decided by the COP26 Governments, to educate and empower people of various professions to contribute to solutions to the climate crisis. This Action for Climate Empowerment (ACE) has the following elements: climate education and public awareness, training, public access to information, public participation, and international cooperation on these matters.

3.7 DEFORESTATION

All people depend upon forests, some more than others. Around three quarters of accessible freshwater in the world comes from forested watersheds. Forests also have a key role to prevent climate related food insecurity. Forests can facilitate the formation of dew from rising water vapour and the accumulation of water in reservoirs and creeks. Forests intercept rain and enhance the water storage capacity of the soil. They also help to conserve the soil, reducing erosion.

Five countries have more than half the forests in the world: Brazil, Canada, China, Russia, and the USA. Almost one third of the global land area is covered by forests and about half the forest area is relatively intact. About one third of all forests are primary forests, where there are no clearly visible indications of human activity.

As noted in 3.6 ceasing deforestation was one of the pledges at the COP26 meeting. The decade 2011–2020 was declared the UN Decade on Biodiversity and some of the conclusions of this work are summarized in an FAO-UNEP report.[78] The impact of deforestation[79] is widely recognized and later in the book we will illustrate some of the couplings to not only climate but to water, energy, food, and nature, as outlined in Figure 3.2. Obviously human health is closely related to most of these aspects.

In the last 30 years around 4.2 million km^2 – corresponding to half the land area of Brazil – have been lost through conversion to other land uses. The global primary forest area has decreased by more than 0.8 million km^2 in the same period and 1 million km^2 have been impaired by forest fires, pests,

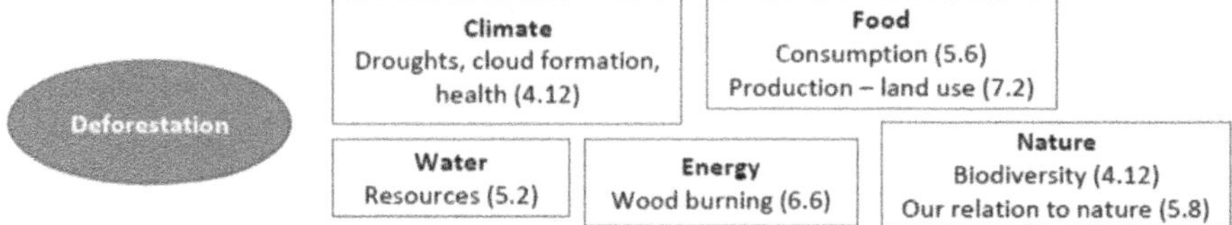

Figure 3.2 The impact of deforestation is further discussed in the sections indicated.

diseases, invasive species, and drought. All of this contributes to the loss of biodiversity. The net deforestation area is lower, 1.77 million km², since new forests have grown by natural expansion or deliberate efforts. As Figure 3.3 shows the rate of deforestation has decreased. The *accumulated* losses for the 3 decades are 4.2 million km² and new forests are 2.4 million km², making the total *net* loss 1.77 million km². The four countries with the highest deforestation have lost 0.44 million km².

Agriculture expansion is the main cause of deforestation and the accompanying loss of biodiversity. In general, the need to provide food and energy for a growing global population is a main cause of deforestation and decreasing biodiversity.

> The *net* deforestation in 30 years = 1.77 million km²

- *Brazil*: Despite the pledge at the COP26 meeting, deforestation in Brazil in 2021 was the highest since 2006: 13 235 km². Expressed differently, a forest area of 1.2×1.2 km² was lost *every hour*, around the clock. During January 2022 the Brazilian Amazon lost a rainforest area of 430 km², five times larger than in January 2021, the highest January forest loss since the records began in 2015,[80] according to Brazil's space agency

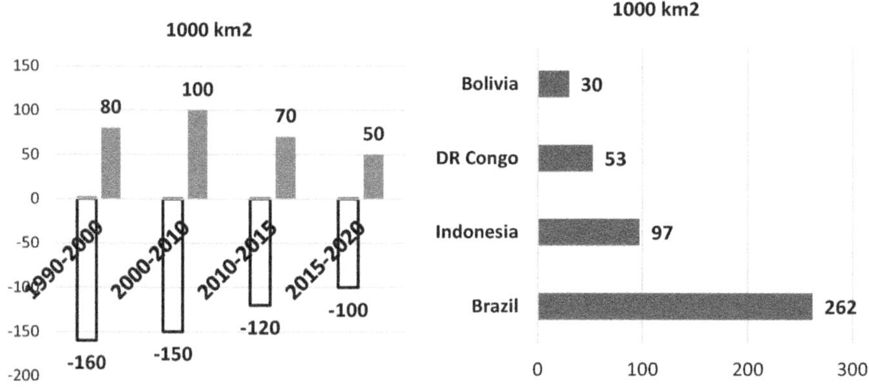

Figure 3.3 The *annual* losses and new forests for three decades (a). The four countries with the highest loss of forests (b).

Inpe. Environmentalists accuse Brazil's President Jair Bolsonaro of allowing deforestation to accelerate. It is suggested that as much as 94% of deforestation in Brazil could be illegal. Deforestation has led to large amounts of habitat and biodiversity loss in the Amazonas. Since 2001, around 100 000–190 000 km^2 of Amazon rainforest has been impacted by fires, potentially affecting the ranges of 77–85% of species that are listed as threatened in this region.[81]

- *DR Congo*: The Congo rainforest is the second largest in the world. Greenpeace estimates that illegal logging is a large cause of the deforestation.[82] Timber gets smuggled outside Congo even though both the USA and EU have banned importing illegal timber. In 2021 Congo lost almost 5 000 km^2 of tropical forest.
- *Indonesia*: the main contributor to deforestation is clearing land for oil palm plantations.
- *Bolivia* lost almost 3 000 km^2 of tropical forest in 2021.

Forests are cleared for local agriculture subsistence. The sad consequence is that the loss of biodiversity will decrease the food supply systems and the ability to adapt to future climate change.

As mentioned earlier the OCO-2 and OCO-3 satellites have provided huge amounts of CO_2 observations. We have always considered the tropical rain forests of the world as the 'lungs of the planet', the most important absorber and storage place of CO_2.

Of global forest cover:

- Tropical and subtropical forests make up 56%
- Forests in temperate regions make up 16%
- Boreal forests in the north make up 27%.

All have an important role to play.

More than 5 years of data from the OCO-2 tell the discouraging story that the tropical regions are a net *source* of CO_2, at least since 2009, according to NASA.[83] Measurements of CO_2 in the tropics are consistently higher than anything around them, and this is still confusing scientists. By observing solar-induced fluorescence (SIF) from chlorophyll in plants, the rate at which plants convert light from the sun and CO_2 from the atmosphere into chemical energy can be measured. It is consistently found that plant respiration is outstripping vegetation's ability to absorb CO_2. It happens throughout the tropics, and almost all the time

However, there are other remarkable consequences. Northern mid- and high-latitude rainforests are now absorbing more CO_2, probably because the growing seasons has become longer. Data so far cannot verify if this has always been the case, but satellite data will produce more evidence with time.

On top of the overshadowing issue of biodiversity, there are so many apparent benefits of trees:

- They reduce heat by providing shade and cool surface temperatures. Urban areas can save air conditioning energy on hot days.
- They absorb carbon and remove pollutants from the atmosphere.

- Trees can improve water quality by taking in polluted surface water and absorbing nitrogen and phosphorus into the soil. Trees reduce flooding by absorbing water and reducing runoff into streams, thus reducing flooding vulnerability (see section 4.3). Trees reduce unpleasant noise while providing enjoyable rustling.

3.8 ACTIONS NEEDED

The needed actions are so apparent and so complex: burning fossil fuels must be phased out as soon as possible. Actions needed are presented in numerous publications, and IPCC has exhaustive lists. They should involve almost all aspects of our lives. Let us just mention a few of them:

- Emphasize the importance of cross-sectoral interactions
- Consider every aspect of our life, where we burn fossil fuels
- How we produce and consume electrical energy
- How we transport goods and people
- How we heat and cool and homes and buildings
- How we take care of and use our water resources
- How we produce and distribute food
- What we eat
- Our economic system and its impact on resource utilization
- How rich countries treat poor countries
- How economic inequality should decrease.

doi: 10.2166/9781789062908_0035

Chapter 4
Global warming impacts

Climate change is already forcing people from their land and homes and putting many more at risk of displacement in the future. Supercharged storms, more intense droughts, rising seas, and other impacts of climate change all magnify existing vulnerabilities and the likelihood of displacement, disproportionately affecting low-income countries, women, children and indigenous peoples.

Extreme heat:
Farmers feel that the environment has shifted from being 'stable' and instead has begun to 'fail the people'. *'The temperature has changed greatly. Summer never used to be so hot, but this year we can hardly work in the fields in the morning. It is so hot that we get blisters all over our body'.* Seken Ali, farmer, Pangsha, Rajbari, Bangladesh.

Drought:
In Ethiopia's Somali region, severe drought has forced pastoralists, including Ibado and her family, to move into temporary settlements. *'In my lifetime I have never seen this. We used to have 700 sheep and goats. Now we have seven. This drought is affecting everyone.'* Ibado, Somali region, Ethiopia.

Sea level rise:
The world's atoll nations, including Kiribati, Tuvalu, and the Marshall Islands, face a truly existential threat from sea-level rise. *'My great hope for my country is that it remains, existing on the map.'* Claire Anterea, Kiribati.

Floods and landslides:
In just a single year deadly floods and landslides have forced 12 million people from their homes in India, Nepal and Bangladesh. The region's monsoon rains are being intensified by rising sea surface temperatures in South Asia. *'We have nothing to survive on. Everything has gone to the stomach of the sea.'* Moneja Begum, Kutubdia Island, Bangladesh.

Moina crossing flood water on her way to her home, Gabura, Shamnagar, West Bengal.

Permission from Oxfam International (www.oxfam.org) *and the Oxfam report 'From Uprooted by climate change' gratefully acknowledged.*

(Photo: Fabeha Monir/Oxfam)

IPCC emphasizes that many of the changes in the climate system become larger as a direct consequence of global warming. This includes both frequency and intensity of hot extremes, marine heatwaves, heavy precipitation, and, in some regions, agricultural and ecological droughts; an increase in the proportion of intense tropical cyclones; and reductions in Arctic Sea ice, snow cover, and permafrost.

It is crucial to remember cause and effect:
Greenhouse gas emissions cause global warming.
Global warming causes climate change.
Climate change has many consequences, including extreme weather

If 'climate change' is used as a synonym for 'global warming' it is tempting for many deniers to say that the 'climate has always been changing'. Global warming is measured reliably.

Feedback phenomena are crucial parts of the climate consideration, as will be described in this chapter. Figure 4.1 illustrates some of the principal paths of actions and phenomena taking place as a result of emission increases and global climate change. Some of these feedback loops will create so called tipping points. Temperature changes will influence the jet stream, cause permafrost and glacier loss, and release methane. Droughts have a profound impact on food production, wildfires, and air pollution. Increased precipitation, storms, and cyclones are closely linked to monsoon patterns and flood disasters. Several of these phenomena are fed back into the climate, strengthening climate change.

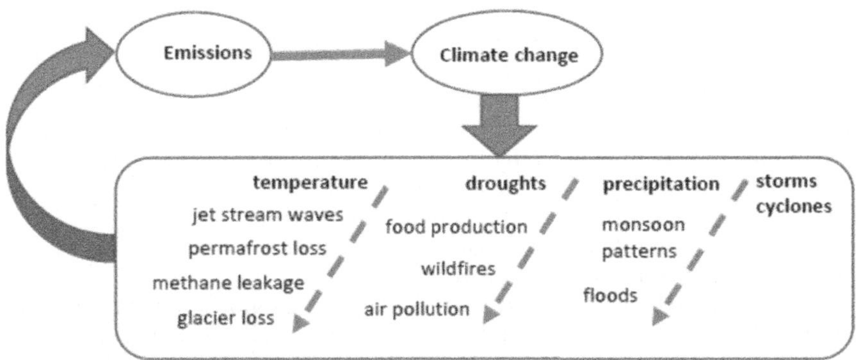

Figure 4.1 Illustration of key feedback loops in the climate system.

4.1 GLOBAL IMPACT OF CLIMATE CHANGE

The USEPA had prepared an important report on the impacts of climate change[84] before January 2017. It stated that 'countries around the world will likely face climate change impacts that affect a wide variety of sectors, from water resources to human health to ecosystems'. The impacts will vary by region. The report also emphasizes that many people in low-income countries are more vulnerable to climate change impacts than people in high-income countries.

It is evident that global warming has consequences for almost everybody. Researchers paired the analysis of more than 100 000 events that could be linked to global warming with a well-established dataset of temperature and precipitation shifts caused by fossil fuel use and other sources of carbon emissions.[85] By using machine learning the findings could connect both to population and land area. The findings focused on events like crop failures, floods, and heat waves and could link escalating extremes and human activities. They concluded that global warming has affected 80% of the world's land area. Weather patterns made worse by climate change have affected at least 85% of the world population. According to one of the co-authors the 85% is 'probably an underestimation.' The study looked at average temperature and precipitation changes, rather than the most extreme impacts, for which there is even more evidence of climate change's role. The machine learning approach has made it possible to identify climate impacts even in places where few studies are made.

4.2 WEATHER VS CLIMATE

As mentioned in Chapter 3, Klaus Hasselmann already half a century ago increased understanding of how weather and climate relate. Now, climate researchers have detected the 'fingerprint' of human-induced climate change on daily weather patterns at the global scale.[86] For a long time, we have made the distinction between weather and climate. The day-to-day weather variability has been clearly separated from the long-term forced response of climate change. The 'weather versus climate paradigm' has been used to explain phenomena that may be perceived as contradictory, such as cold winter days in a warming climate. That weather is not the same as climate is true locally. So, if we only monitor the weather in one particular place, then we cannot conclude much about climate. However, on a *global scale* this is no longer true. The global mean temperature on a single day is already quite a bit shifted. Therefore, one can see this human fingerprint in any single moment. Quoting the authors: 'weather is climate change if you look over the whole globe'[86.]

> *Weather = climate change* if the whole globe is included

The study concludes that the spatial patterns of global temperature and humidity are, in fact, distinguishable from natural variability and have a human component to them. The fingerprint of climate change is detected from any single day in the observed global record since early 2012, and since 1999 on the basis of a year of data. The results have modified the climate change narrative: while changes in weather locally are emerging over decades, global climate change is now detected instantaneously, and the authors conclude that 'climate is what you expect, weather is what you get'.

4.3 EXTREME WEATHER

Many weather and climate extremes are appearing in every region across the globe. The evidence that human influence has caused the changes in extremes – heatwaves, heavy precipitation, droughts and tropical cyclones – has been strengthened during the last decade.

Climate-related disasters are getting more common. From 2005 to 2015, the UN found 335 weather-related disasters (droughts, storms, and others) each year across the globe, almost twice as many as during the period 1985–1994. The average costs per catastrophe are also increasing. The inflation-adjusted cost of natural disasters was about US$30 billion per year in the 1980s. After 2015 the annual cost has increased six-fold to an average of US$182 billion. As an example of connections between global warming and extreme weather, the deadly North American Pacific heat wave in June 2021 (see below) would have been 'virtually impossible' without greenhouse gas pollution according to recent research.[87]

Storms have resulted in the largest economic losses around the world. In the 1970s the global losses were US$49 million/day or US$18 billion/year. This increased more than seven times to US$383 million/day or US$140 billion/year in the 2010s.

> Global losses from storms have increased from US$18 to 140 billion/year in 50 years.

A report from Christian Aid[88] has summarized the costs of the most serious events in 2020, related to climate change. The events were not only causing devastating financial costs but were catastrophic for millions of people. The estimated costs are based on insurance information and the costs are usually higher in richer countries. The true figure is most likely much higher, in particular in low-income countries, where extreme events have been overwhelming, even if the financial cost is not shown adequately.

If the disaster costs in a country are related to the average income of a person in that country, another picture will appear. The costs in the various countries, shown in Figure 4.2, have been divided by the gross domestic product (GDP) per capita, often considered an indicator of a country's standard of living. It is not a measure of personal income. When the costs are normalized with the GDP the relationship between the various disasters becomes dramatically different. The suffering of an individual in a rich country may be as large as it is for a person in a poor country. However, the ability to mitigate damage in rich countries is dramatically demonstrated in Figure 4.3. The US West Coast and the Australian fires as well as the European windstorms got a lot of media attention in the West, while the devastating locust invasions in East Africa were hardly mentioned in most western media.

Fires and floods look like logical consequences of climate change. The locust invasion in East Africa in early 2020 may need more explanation and was caused by an unusual rainy season at the end of 2019. Locusts thrive in wet conditions and in 2020 they invaded large regions, destroying huge areas of crops and trees, almost like the Biblical story from Exodus chapter 10, verses 14–15. The cause of the unusual weather patterns is related to the Indian Ocean Dipole (IOD), maybe less known than El Niño in the Pacific. IOD is an ocean circulation pattern that affects the climate in the region. A positive IOD means a wetter and warmer than normal western ocean (=East Africa) and a cooler than average conditions in the east (including western Australia). The 2019 positive IOD was the strongest in 60 years. It is predicted that climate change will make the positive IOD stronger and more frequent. This will have devastating effects in already vulnerable regions.

In South Sudan more than 1 million people were affected in 2020 by huge rains and overflowing rivers. The Blue Nile River rose more than 17.5 metres, a new record. In the Pantanal in the Amazonas, the world's largest wetland (210 000 km^2), fires affected more than 45 000 km^2. Nearly 17 million animals died in the Pantanal fires in 2020[89].

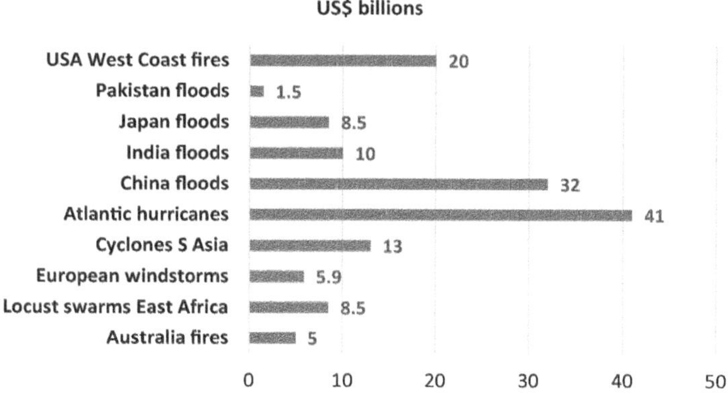

Figure 4.2 The most expensive weather-related disasters in 2020 shown in billions of US$. The human suffering is only partially reflected in these figures (Source: Christian Aid).

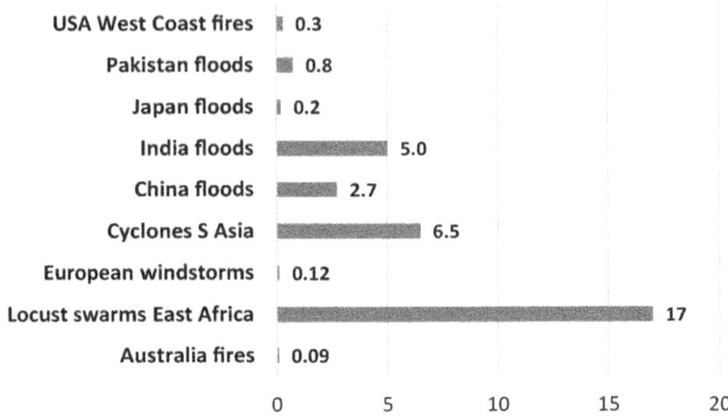

Figure 4.3 The weather-related disaster damage costs, shown in Figure 4.2, normalized with the national GDP/capita. The Atlantic hurricanes are not considered since they affected several countries. (GDP source, International Monetary Fund, 2021).

WMO and the UN Office for Disaster Risk Reduction (UNDRR) have reported that climate change and increasingly extreme weather events have caused a surge in natural disasters during the last half century (see also Figure 4.4). This has impacted poorer countries disproportionately. Natural hazards accounted for half of all the disasters, 45% of all reported deaths, and 74% of all economic losses. There has been 11 000 weather, climate, and water-related disasters globally since 1970, causing over 2 million deaths. More than 91% of the deaths occurred in low-income countries. However, the good part of the news is that the number of deaths has decreased thanks to improved early warning systems and disaster management, a decrease from 50 000 in the 1970s to less than 20 000 in the 2010s.

Figure 4.4 reflects the great differences between low- and high-income countries. While all of Africa showed economic losses of only 1%, North

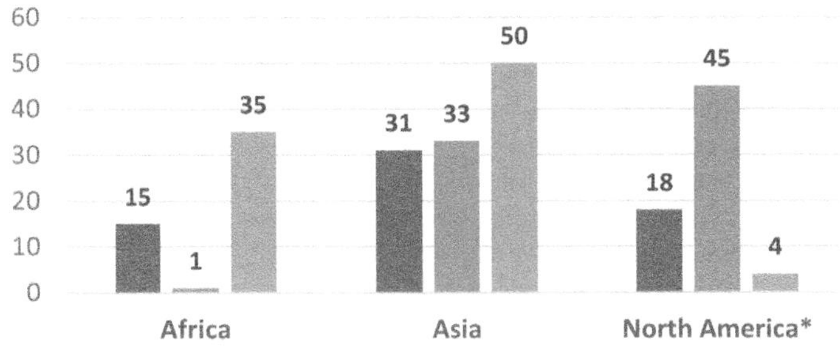

Figure 4.4 Weather, climate, and water-related disasters between 1970 and 2019 on three continents.[90] Left bars: fraction of the number of global disasters (in %). Middle bars: fraction of global economic losses for the disasters (in %). Right bars: percentage of the numbers of deaths. *Including Central America and the Caribbean.

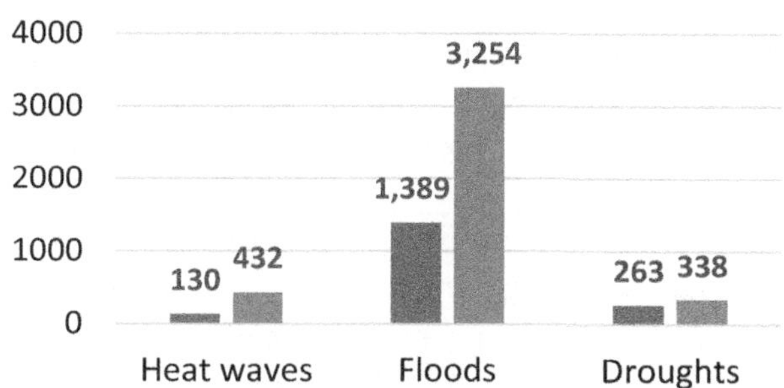

Figure 4.5 Number of heat waves, floods, and droughts during 1980–1999 (left bar) and during 2000–2019 (right bar).[91] Note: The events have the following criteria: more than 10 deaths, and/or at least 100 injured or homeless.

America had 45% of the costs. The USA alone had 38% of the global economic consequences of the disasters. The difference in the numbers of deaths may demonstrate that high-income countries have better protection in combination with better warning systems.

The increase of extreme weather events is illustrated in Figure 4.5.

The dominating cause of death due to disasters is different in various regions of the world. Table 4.1 shows the most common disaster and its death toll in the different continents.

IPCC writes about *compound* extreme events. This includes increases in the frequency of concurrent heatwaves and droughts on the global scale, fire weather in some regions of all inhabited continents, and compound flooding in some locations. Human influence has likely increased the probability of such compound extreme events since the 1950s. IPCC concludes that the global proportion of category 3–5 tropical cyclones has increased since the 1980s. Physical understanding indicates that human-induced climate change increases heavy precipitation associated with tropical cyclones.

Table 4.1 The dominating type of extreme weather in the different continents and its death toll in % of the total deaths caused by weather, climate, and water-related disasters.

Continent	Responsible for most deaths	Fraction of all deaths (%)
Africa	Drought	95
Asia	Storms	72
South America	Floods	77
North and Central America + Caribbean	Storms	71
Southwest Pacific	Storms	71
Europe	Extreme temperatures	93

Source: UNDRR.

Table 4.2 Frequency of extreme temperature events, heavy one-day precipitation events, and intensity of agricultural and ecological drought events that would occur in 10 or in 50 years on average.

	1850–1900	Present	Future global warming		
	0°C	1°C	1.5°C	2°C	4°C
High temperature extremes over land					
10-year event	1	2.8	4.1	5.6	9.4
50-year event	1	4.8	8.6	13.9	39.2
Heavy precipitation over land					
10-year event	1	1.3	1.5	1.7	2.7
Agricultural & ecological droughts in drying regions					
10-year event	1	1.7	2.0	2.4	4.1

The period 1850–1900 represents a climate without human influence. The intensity of the temperature extremes, precipitation and droughts are also expected to increase with global warming (Source: IPCC, 2021).

The probability for extreme events will increase with temperature. The definition of a 10-year event is that one event – on average – will occur in 10 years. A similar definition is, of course, valid for other time periods. Table 4.2 gives the IPCC estimate how the probability for events depends on global warming.

Given the pledges at the COP26 meeting we would expect one 'high temperature 10-year event' every 1.5–2 years. Similarly, the world would experience a heavy precipitation or serious drought '10-year event' every 5 years.

4.3.1 Human-made vulnerability

It is easy to blame all disasters on climate change. This is not true. Natural hazards such as floods, droughts, and heatwaves are related to climate but become disasters also because of societal vulnerability.[92] This is often constructed. There have been devastating results of irresponsible irrigation. Vulnerability is therefore a product of social and political processes that include elements of power and (poor) governance.

Climate change was blamed for the floods in Germany in 2021. However, vulnerability increased the damage. Hydrologists monitoring the river flows note that farmland has increased on the once-boggy hills where the rainfall was most intense. Before the land was more sponge-like and could more easily absorb heavy rains. Removal of natural vegetation, field drains, and roadways made it possible for the water to reach the river much faster. Therefore, restoring the former sponges could reduce the peak river flows. Fighting climate change takes much longer.

4.4 TEMPERATURE

IPCC reports that hot extremes (including heatwaves) have become more frequent and more intense across 41 out of 45 land regions in the world since the 1950s. Some recent hot extremes observed over the past decade would have been extremely unlikely to occur without human influence on the climate system. Cold extremes (including cold spells) have become less frequent and less severe. Again, human-induced climate change is the main driver of these changes.

A phenomenon called a *heat dome* has caused record heatwaves that have been both longer and more intense. During still and dry summer conditions a mass of warm air builds up and rises. Then an extreme high pressure in the atmosphere will push down the hot air, causing the air to become compressed and trapped in place. It may cause temperatures to soar over an entire continent. The heat dome phenomenon is found to have caused the record heatwaves in Western Canada and the USA during 2021. Normally a heat dome is expected to take place once in a thousand years.

The record temperature in Lytton, western Canada, of 49.6°C was a wake-up call. The reason is not that a new temperature record was set. This happens from time to time. The previous all-time Canada record of 45°C was set in 1937. Normally records are exceeded by a fraction of a degree. This time the final temperature in Lytton was 4.6°C higher than the old record. A heatwave in Siberia in 2020 saw temperatures more than 5°C above the previous record between January and June. The findings are exactly what we should expect, according to climate research.[93] Unlike other weather extremes, heatwaves do not leave a trail of destruction; they are 'silent killers'. People die quietly in poorly insulated homes or houses without air conditioning. For example, the European heatwaves in the summer of 2003 were responsible for the deaths of tens of thousands of people.[94] The 2010 Russian heatwaves killed an estimated 55 000 people.[95] It is estimated that 37% of deaths caused by heatwaves are related to climate change caused by human activities.[96]

> Heatwaves are 'silent killers'

We typically learn about standard temperature measurements. In extreme conditions it is more important to talk about wet-bulb temperature (TW) that accounts for both heat and humidity. The TW reflects what this combination means for the human body's ability to cool down. Exposure to extreme heat can lead to heat stroke and dehydration, as well as cardiovascular, respiratory, and cerebrovascular disease. Excessive heat is more likely to affect populations in northern latitudes where people are less

> The wet-bulb temperature determines the human body's ability to cool down.

prepared to cope with excessive temperatures. Dry heat feels more tolerable than extreme humidity. This is well-known for anybody having experienced a sauna bath. If the TW reading is higher than our body temperature, it means that we cannot cool ourselves to a temperature tolerable for humans by evaporating sweat, and that basically means we can't survive. A TW of 35°C marks our upper physiological limit. Some coastal subtropical locations have already reported a TW of 35°C. This extreme humid heat overall has more than doubled in frequency since 1979.[97] Humid heat is increasingly severe.

During 2020 several temperature extremes were reported with 54.4°C in Death Valley, California, 39.7°C in Cuba and 38°C in Verchojansk in Siberia. The last record is the highest temperature ever recorded in the Arctic. It was officially confirmed on 14 December, 2021, by the World Meteorological Organization (WMO). The temperature was 18°C higher than the area's average daily maximum for June.

The average temperature in Siberia was 3°C higher than normal, leading to extensive fires. The 2021 summer in the Russian republic of Sakha in north-eastern Siberia has been the driest summer in 150 years. Almost 190 fires ravaged the region, over an area of 34 000 km^2, an area larger than Belgium. Most of it is forest land. As a sign of the extent of the fire, NASA reported in August 2021 that smoke from the fires had reached the North Pole, some 3 000 km from Sakha. The European surveillance satellite Copernicus reported that the fires had generated more than 500 Mt (or 0.5 Gt) of CO_{2e} during the three summer months of 2021.

As a result of permafrost melting, a corroded fuel storage tank collapsed in May 2020 in Norilsk, Krasnoyarsk Krai, in the north of Russia. The temperature in the region had risen far above normal in the days before the disaster. Norilsk Nickel is the world's largest producer of refined nickel, a metal that is not only critically needed in stainless steel, but also in batteries for electric cars. Around 17 500 tons of diesel oil leaked from the storage tank into the Daldykan River and reached the Kara Sea in the Arctic Ocean. The river had already been severely polluted several years earlier and had turned red from wastewater, coming from the smelting plant. On top of the river pollution Norilsk Nickel is one of the largest air polluters in Russia, emitting 1.9 Mt of sulfur dioxide (SO_2) per year, averaged over the last 16 years. The number two SO_2 emitter was the eruption of the Mount Etna volcano, Sicily, Italy, with 1.1 Mt. There are, on average, 45–50 volcanic eruptions happening at the same time on Earth. The eruptions certainly attract attention, but they are dwarfs compared to human-made emissions. Three days of human emissions are equal to the entire CO_2 emissions from all volcanoes on Earth in a whole year.

Another worry concerning thawing permafrost is in Alaska. The thaw threatens to undermine the supports holding up an elevated section of the Trans-Alaska Pipeline. This is one of the largest oil pipelines in the world. If the structure of the pipeline is jeopardized, there is a risk of oil spill in a remote region where it would be extremely difficult to clean up.

The Arctic is fragile, and the future may well be summed up as: more rain, less snow. Recent research has found that the shift could come decades sooner than previously thought[98] and parts of the Arctic could become dominated by rain rather than snow already within 40 years, particularly if the Earth continues to warm at its current rate. Still there are uncertainties and the climatologists note that 'precipitation is one of the most difficult variables for models to get right.'

Also, from the southern polar areas some disturbing temperature records have been reported. WMO reported a new record for Antarctica on 6 February, 2020: 18.6°C. During the last 50 years the Antarctica yearly average has become 3°C warmer. Also, Sweden has become 3°C warmer compared to pre-industrial times.

Global climate data are usually presented in gridded square 'locations' that make up an area of about 25 km². In the 1980s, 220 locations had recorded more than 50°C at some point during the year. This number has increased every decade since then. During the last decade 876 locations exceeded 50°C.[99] Higher temperatures will have a major influence on how people manage to work outdoors. It is estimated that regions in South Asia, Africa, and Central and South America may lose up to 250 working days per year in the year 2100.

4.5 ROSSBY WAVES

Another phenomenon appears when a storm fuelled by warmer-than-normal sea temperatures disrupts the jet stream. The jet stream is a current of fast-flowing air moving from west to east high up in the atmosphere. When a storm distorts the stream, it behaves like yanking a long skipping rope at one end and seeing the ripples transferring along it. These are the Rossby waves, described by the Swedish meteorologist Carl-Gustaf Rossby. Already in 1955, by using a computer, he had managed to predict atmospheric grand currents 2–3 days ahead. The waves are quite common in the summer months in the Northern Hemisphere, but recently they have been extreme. When that happens, warm air travels further north and cold air penetrates further south. The result is a succession of unusually hot and cold weather systems along the same latitude. The jet stream waves cause everything to slow down, and weather systems can remain over the same areas for days on end. There can be heatwaves that seem to last forever or extreme rains.

So, the question is if the Rossby waves are caused by the global warming. Climate research points to the Arctic area conditions. Since the temperature increase in the Arctic is higher than regions further south the temperature difference between the polar region and the tropics decreases. The decreasing temperature difference will slow down the jet stream in the upper part of the atmosphere. The Rossby waves explain not only the heatwaves in the summer but also extremely cold winter conditions. For example, in February 2021:

> *A sign of the complexity*: global warming can cause extremely cold weather

- *Texas*: there was extreme cold in Texas (Dallas had record low temperature of −19°C). 3.5 million businesses and homes were left without power.
- *Sweden* had a new record high temperature in February, 17.0°C. Likewise Poland had record high 22.1°C and Slovakia 20.8°C.
- *Amsterdam, The Netherlands*: people could skate on the canals
- *Athens, Greece*: the Akropolis was covered by snow.
- *Beijing* in China was record warm with 25.6°C in the middle of winter, 6°C warmer than the previous January/February record.

The jet stream can bend towards south and allow cold air flow all the way to southern USA or southern Europe. Similarly warm air can travel northwards below the 'tops' of the Rossby waves and cause extremely warm winter temperatures.

The climate models did not predict this kind of weather for a 1.2°C warming, so apparently there are more driving forces causing extreme weather. It has been expected that the increase of extreme weather will increase proportionally to the temperature increase. However, there has been several 'outliers' that seem to depend on more factors than only temperature. There are still uncertainties among climatologists how the Rossby waves will influence weather patterns. They simply will need longer measurement records for the climate models.

Today, researchers are focusing their efforts on predicting swings in the jet stream. Most research has focused on the Northern Hemisphere. Impacts in the south, particularly in South America and the southern tip of Africa, are less well understood because there's less research and raw data from that part of the world.

4.6 DROUGHTS

IPCC reports that human-induced climate change has contributed to increases in agricultural and ecological droughts in 12 out of the 45 global regions due to increased land evapotranspiration, notably in Western North America, Western and Central Europe, around the Mediterranean, most African regions, Western, Central and East Asia, and in Western Australia.

Droughts can worsen as heatwaves become more intense and longer. Soil moisture and water supplies run dry more quickly causing the ground to heat up quicker, warming the air above and leading to more intense heat. Simply expressed: drought begets drought. When the soil is dry, the sun's energy is focused on heating the air instead of evaporating water. That increases temperatures, leading to more dryness, which allows drought to spread even further.

Dry conditions increase the possibility for fires. The frequency of large wildfires has increased dramatically in recent decades. Intense wildfires reach temperatures above 800°C and can essentially create their own weather systems, forming so called *pyrocumulonimbus* clouds, thunder clouds that can produce lightning and ignite more fires. These fires can spread at an incredible speed, as experienced, for example, in Australia, in Western Canada, in California, in Greece, in Siberia, and even in Sweden, to mention some recent events. In California it is estimated that areas hit by fires during the summer months June–September has increased eight-fold since the 1970s.

A quarter of the world's major cities face a situation of water stress. A 2014 survey of the world's 500 largest cities estimated that one in four are in a situation of 'water stress'. In December 2017 I visited Cape Town, in South Africa. The city had been affected by a serious drought for years and was in the unenviable situation of being the first major city in the modern era to face the threat of running out of drinking water. During my visit there were the first mentions of plans for 'Day Zero', a shorthand reference for the day when the water level of the major dams supplying the City would fall below 13.5%. Nearby, the Eastern Cape drought has continued in 2021.

> One in four
> of the world's
> major cities
> face a situation
> of water stress.

The Cape drought was analysed using five different large model ensembles as well as observed data. The results indicated an increase of the probability by a factor of three of such a drought occurring because of anthropogenic climate change. All the model results further suggest that this trend will continue with future global warming. These results are in line with physical understanding of the effect of climate change at these latitudes and highlights and motivates those measures to improve Cape Town's resilience to future droughts are an adaptation priority.[100]

African nations that make the smallest contribution to global warming are hit by acute droughts. In February 2022, the Horn of Africa, including Ethiopia, Kenya, and parts of Somalia, has around 13 million people facing severe hunger amid the driest conditions since 1981, according to the UN World Food Programme. This is the third consecutive failed rainy season. Most water sources that have usually been resilient to climate variability have dried up in Kenya. In Kenya alone, the drought killed 1.4 million livestock in the final part of 2021. In southern Ethiopia, about 0.24 million livestock died.

Several major cities in the world face similar water supply crises, caused by a combination of high consumption and too little precipitation combined with too much pollution. São Paulo, Brazil faced a dangerous situation in 2015, when the main reservoir fell below 4% of the full volume. Bangalore, India, struggles with water pollution. An inventory of the city's lakes found that 85% had water that could only be used for irrigation and industrial cooling.[101]

4.7 AIR POLLUTION FROM WILDFIRES

It is obvious that all the elements of our climate are connected. It takes more than a spark to start a wildfire. There has been an increasing wildfire activity in combination with global increases in temperature, drought, and extreme weather. Climate change also affects lightning strikes, another important ignition source. The resilience of forests is the capacity of the forest to return to the state before the disturbance, be it a drought or a fire. Temperature and drought have the largest impact on trees in their youngest life stages, and forest resilience to disturbances due to climate change conditions remains highly uncertain.

Between 2003 and 2012 around 670 000 km² of forest burned around the globe *every year*. The year 2015 was particularly devastating when 980 000 km² of the world's forest burned. Most of the fires in 2015 occurred in the tropical domain, where around 4% of the total forest area was consumed. Most of the affected forests were in Africa and South America.

> An area of the size of France is destroyed every year from forest fires.

There are clear indications that the risk of wildfires will continue to increase in most areas of the world as climate change worsens. The devastating fires in Siberia were mentioned in 4.4. During the last 2–3 years, there has been an immense scale of wildfires in Australia, the Amazon rainforest in Brazil, the western USA, and British Columbia, Canada. From 1997 to 2016, the global mean CO_2 emissions from wildfires equated to approximately 22% of the carbon emissions from burning fossil fuels[102].

The fires will increase excess mortality and morbidity from burns, wildfire smoke, and mental health effects. Worsened air quality that often accompanies heat waves or wildfires can lead to breathing problems and exacerbate respiratory and cardiovascular diseases[103]. Substantial greenhouse-gas emissions and forest loss from wildfires are likely to accelerate climate change further and possibly lead to a reinforcing feedback loop.

Wildfire smoke contains several components: particulate matter (PM), carbon monoxide (CO), nitrogen oxides (NO_x), such as nitrogen dioxide (NO_2) and nitric oxide (NO), and volatile organic compounds. A photochemical reaction under sunlight between volatile organic compounds and nitrogen oxides produces ground-level ozone, a secondary pollutant. Too much ozone exposure makes it more difficult to breathe deeply or make the lungs more susceptible to infection. It can increase the frequency of asthma attacks. Wildfire PM tends to have a smaller particle size than PM from fossil fuel combustion. It also contains oxidative components (e.g., aromatic hydrocarbons and oxides of nitrogen) that potentially lead to stronger toxic effects.

An increasing global temperature is likely to reinforce wildfires. They will in turn create a feedback loop back to climate change and worsen the consequences. Wildfires and fossil fuel burning are appalling sources of air pollution and threaten public health.

> In 2019, forest fires produced >1.8 Gt of CO_2, ~ the emission rate of international transportation

4.8 PRECIPITATION

IPCC states that heavy precipitation events will intensify and become more frequent in most regions with additional global warming. The frequency and intensity of heavy precipitation events have increased since the 1950s over most land area for which observational data are sufficient for trend analysis (high confidence), and human-induced climate change is likely the main driver. At the global scale, extreme daily precipitation events are projected to intensify by about 7% for each 1°C of global warming.

Warmer weather can cause not only dryer conditions but also more extreme rainfall events. In the usual weather cycle, hot weather leads to the build-up of moisture and water vapour in the air, which turns into droplets to produce rain. With increasing temperature there is more vapour in the atmosphere, resulting in more droplets – and heavier rainfall. Often this happens in a shorter time and over a small area. Weather around the globe will always be highly fluctuating but climate change will increase the extreme events. Also, as some regions get dryer, as in Siberia and the western USA, water will fall in other places, like recently in Germany, Belgium, and Japan. In Stockholm, Sweden, the month of May 2021 was the wettest since 1786.

> Warmer weather can cause both dryer conditions and extreme rainfall.

The monsoon precipitation over South Asia, East Asia, and West Africa has increased since the 1980s due to warming from greenhouse gas emissions and counteracted due to cooling from human-caused aerosol emissions.

Increasing precipitation and flooding will have an impact on water supplies and water quality.

Evaporation and evapotranspiration influence the energy balance and temperature at the ground level. Heat is transported via the water vapour from the ground to the atmosphere, causing the ground to be cooler. Since water vapour is also the most important greenhouse gas, it creates positive feedback. A higher air temperature leads to more water vapour that leads to higher temperature. These changes are not only related to annual averages. The variations can be significant from day to day, month to month, or year to year. This in turn will influence the frequency, magnitude, and duration of extreme weather.

4.9 METHANE

After water vapour and CO_2, methane (CH_4) is the most critical greenhouse gas and accounts globally for approximately 18% of global warming. Methane is an odourless and colourless gas with density of 0.6 relative to air. Since it is lighter than air it rises when released.

The concentration in the atmosphere is 1.86 ppm compared to 0.72 ppm during pre-industrial times. If the effect is measured for a full century, then methane has about 60 times higher ability to absorb heat compared to CO_2. Unlike CO_2, methane does not have natural sinks so there is no natural methane cycle. Fortunately, the methane does not remain in the atmosphere as long as the CO_2. The gas tends to decay quite rapidly by oxidizing in the atmosphere, leaving as final products water and CO_2.

> Methane accounts globally for 1/5 of global warming.

Unlike other GHGs, methane can be used to produce energy since it is the major component (95%) of natural gas. Consequently, for many methane sources, opportunities exist to reduce emissions cost-effectively or at low cost by capturing the methane and using it as fuel.

IPCC is quite cautious about mentioning feedback from methane leakages but note that '*additional ecosystem responses to warming not yet fully included in climate models, such as CO_2 and CH_4 fluxes from wetlands, permafrost thaw and wildfires, would further increase concentrations of these gases in the atmosphere.*' Far from declining, 2020 saw methane emissions grow at the fastest rate in 40 years. The increased use of natural gas to replace coal may cause further methane concentration increase. This will also create increasing risks for leaking gas.

Natural sources and sinks of non-CO_2 greenhouse gases such as methane and nitrous oxide (N_2O) respond both directly and indirectly to atmospheric CO_2 concentration and climate change, and thereby give rise to additional biogeochemical feedbacks in the climate system. Many of these feedbacks are only partially understood and are not yet fully included in earth system models. One of the big climate concerns is that the ongoing global warming may cause a rise in warming gases from natural sources. Scientists are concerned that this is already happening with methane. For example, the 2020 Siberian unusually warm summer caused increasing methane emissions from the permafrost. The emissions are still relatively small, but there are reasons to be attentive.

Around 60% of the methane that ends up in the atmosphere comes from human sources such as agriculture, fossil fuels, landfills and biomass burning. The other 40% comes from microbial activities in natural sources such as wetlands. When precipitation increase in wetland areas and the temperature is rising then the activity of methane-producing microorganisms will increase and produce more methane. The International Energy Agency (IEA) estimates that the oil and gas sector emitted around 70 Mt of methane (approximately 2.1 Gt CO_2-eq) in 2020 – just over 5% of global energy-related greenhouse gas emissions.[104]

There is a huge amount of old abandoned oil wells globally, and just in the USA there are hundreds of thousands.[105] These wells play a crucial role in the climate crisis. They mostly leak natural gas, consisting of some 95% methane. A strange kind of business is developing, where used wells are bought cheaply. The buyer will use and sell the trickle of gas or oil that is left and still make some profit. In the USA, state laws require that every well be plugged with cement after it runs dry,. However, many well owners do not meet the legal requirements and abandon the wells. If the company goes broke, then the burden to tighten the leaks falls to the taxpayers. So far owners of leaking oil and gas wells are not breaking any rules as the wells have not been abandoned, and hitherto there have been no restrictions on methane emissions – until COP26. The cost for quickly fixing leaking wells, for example, in Pennsylvania is remarkably low, less than US$90 on average. Still greed is stronger than environmental responsibility.

> Leaking methane is a significant source of greenhouse gas emissions

Controlling methane leakages is among the cheapest and quickest ways to slow climate change. Nearly half of the roughly 380 million metric tons of methane released by human activities annually could be cut this decade with available and largely cost-effective methods. To combat human-caused methane emissions could prevent as much as one third of the warming expected in the next few decades.[106] Methane emissions from livestock and agriculture pose the challenge that require changes in how farmers raise crops and feed livestock – not to mention behavioural adjustments by consumers fond of their hamburgers. Five areas contribute 90% of all methane emissions, as Figure 4.6 shows.

In a report from Global Energy Monitor,[107] 432 proposed coal mines have been analysed with respect to methane leakages. In a surface coal mine, the methane gas content is estimated to be in the range of 1–3.5 m³ (median = 1.7 m³) per ton of coal. For an underground coal mine the methane gas content is many

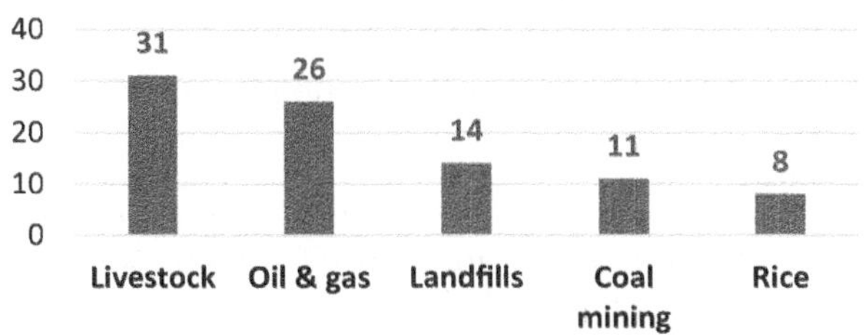

Figure 4.6 Contribution from the five dominating methane sources (%).

times higher, in the range 9–14 m³ (median = 13.7 m³). Some mines can contain 24 m³ per ton.

If not mitigated, these mines – under construction or pre-construction – will leak more than 13 Mt of methane per year, or a 30% increase compared to current emissions. The methane leakage corresponds to more than 1.1 Gt of CO_2-equivalents (CO_{2e}) on a 20-year horizon or 0.4 Gt per year on a 100-year horizon. These findings are confirmed by the IEA,[108] reporting that methane accounted for approximately 9–10% of mine-level emissions on a 100-year horizon in 2018.

The NOAA Annual Greenhouse Gas Index (AGGI) shows that the methane concentration in Earth's atmosphere has been surging over the past half-decade.[109] Despite the pandemic, 2020 saw the biggest one-year jump on record. The causes of the recent spike are unclear, but could include natural gas fracking, increased output from methane-producing microbes spurred by rising temperatures, or a combination of human-caused and natural forces.

Today methane leaks can be detected from space, thanks to the development of imaging technology and methane-hunting satellites. The French Kayrros Methane Watch is a monitoring platform that measures methane footprint on a global scale, using a combination of satellite observations and algorithms. Satellite observations have detected leaks in remote regions, such as leaking pipelines in Siberia and in Kazakhstan. A cluster of satellites launched by national space agencies and private companies over the last 5 years have greatly sharpened our view of what methane is being leaked from where. As recently as 2020 methane-hunting satellites have made several worrying discoveries. For example, methane emissions from oil and gas operations in Russia rose 32% in 2020, despite the pandemic. Satellites also observed extensive releases from gas pipelines in Turkmenistan, a landfill in Bangladesh, a natural gas field in Canada, and coal mines in the US Appalachian Basin, and in Australia.

At any given time, according to Kayrros, there are about 100 high-volume methane leaks around the world. On top of that there is a mass of smaller ones that add significantly to the total. The mini-fridge-sized satellites will be able to target a release to within 30 metres, precise enough to identify the exact piece of equipment that's leaking.

4.10 OTHER SIGNIFICANT GREENHOUSE GASES

There are two gases that require special attention from a climate point of view.

4.10.1 Nitrous oxide

IPCC estimates that N_2O comprises roughly 6% of greenhouse gas emissions and around 75% of those N_2O emissions come from agriculture, typically from the use of nitrogen fertilizers. N_2O is about 300 times as potent as CO_2 at heating the atmosphere. It also depletes the ozone layer. The gas continues to accumulate. A 2020 review of N_2O sources and sinks found that emissions rose 30% in the last four decades.

It has been recognized for a long time that N_2O can be emitted in wastewater treatment operations.[110] N_2O can be emitted during nitrogen removal, even if there are great variations between different plants, depending on the operational strategies. Important parameters are low dissolved oxygen concentration in nitrification and low carbon/nitrogen ratio in denitrification

4.10.2 Hydrofluorocarbons

Hydrofluorocarbons (HFC) are used in refrigeration and air conditioning and often found to be leaking from freezers. HFCs were once an environmental solution – not a problem. They replaced other chemicals that, when released into the atmosphere, eroded the Earth's protective ozone layer. But their heat-trapping properties still exacerbated global warming. Five years ago, the Kigali Amendment[111] was signed to deal with these pollutants. For example, emissions from one by-product called HFC-23 is more than 12 000 times CO_2-equivalents over the course of a century.

The Olympic and Paralympic Games Beijing 2022 used CO_2 refrigeration systems at four ice sports venues. The CO_2 technology is replacing HFCs, traditionally used to cool ice rinks. CO_2 is collected and purified from industrial waste gases, while the waste heat generated during the refrigeration process is recycled and used for the stadium's ambient heating, hot water for ice making and melting and showers, and the control of indoor temperature and humidity.

4.11 CLIMATE FEEDBACK MECHANISMS

There are dangerous positive feedback mechanisms caused by climate change, the so-called tipping points, an idea introduced by IPCC some 20 years ago. They are moments at which the dire effects of global warming will become irreversible. The mechanisms are nonlinear in the sense that a small change in forcing triggers a strong response in some part of the climate system.[112] This serious risk has been widely described in the scientific literature during the last decade. IPCC claims that tipping points could be exceeded even for warming in the range 1–2°C.[113] They include phenomena like:

- *Irreversible melting of the Greenland ice sheet*: the Greenland ice sheet is melting at an increasing rate. If it passes a particular threshold, it could add another 7 metres of the sea level rise over thousands of years. Climate models suggest that the Greenland ice sheet could be doomed at 1.5°C of warming, which could happen as soon as 2030.

- *Arctic Sea ice*: this is already shrinking rapidly. At 2°C of warming, the region has a 10–35% risk of becoming largely ice-free in summer. The shrinking of Arctic Sea ice is a result of global warming, but it is also causing the planet to warm quicker. More sunlight is being absorbed by the darker ocean, rather than being reflected into space by the so-called albedo effect. Sea ice will reflect some 80% of sunlight back into space, while the darker ocean surface will absorb about 90% of the sunlight hitting it. Data from NASA suggest the loss of the minimum Arctic sea ice extent is of the order of 13% per decade, based on the 1981–2010 average.
- *The West Antarctic ice sheet*: the Amundsen Sea embayment of West Antarctica might have passed a tipping point: the 'grounding line' where ice, ocean, and bedrock meet is retreating irreversibly. There is a risk that cannot be neglected, that when this sector collapses, it could destabilize the rest of the West Antarctic ice sheet like toppling dominoes. This could lead to about 3 metres of sea-level rise. The rate of melting depends on the magnitude of warming above the tipping point. At 1.5°C, it could take 10 000 years to unfold; above 2°C it could take less than 1000 years.
- *Amazon rain forest* is not only the largest rain forest in the world, but also home to 10% of all known species. The tipping point in Amazonas could be anywhere between 20 and 40% deforestation. Since 1970, 17% of the rainforest has been lost. Due to deforestation and climate change, there are indications that the rain forest would be eroded past the point at which it can produce enough rainfall to sustain itself. The impact is clear: less forest means less rain. Then more forest would die, causing still more rainfall decrease.
- *Permafrost*: across the Arctic there are worrying signs of irreversible thaw, causing the release of both CO_2 and of methane. Permafrost emissions could take an estimated 20% off the remaining emissions budget to staying within 1.5°C increase.
- *Atlantic circulation*: the Atlantic Meridional Overturning Circulation (AMOC), also known as Gulf Stream System, has been weaker in the last decades than any time during the last 1000 years.[114] The giant ocean circulation is relevant for weather patterns in Europe and regional sea-levels in the USA. The system is the ocean's thermohaline circulation ('thermo' = heat, 'haline' = salt), and it plays many critical roles in the climate. The Gulf Stream is also called the global ocean conveyor belt because it redistributes heat worldwide.

Note, that the tipping points do not act as isolated phenomena. They may happen in cascade. Having reached one tipping point may make it more probable the next one will be reached. What is hardly recognized in most public discussions on climate change is that cascading tipping points are an existential threat to not only humanity but to life at Earth. Then no amount of economic cost–benefit analysis is going to help us. The world needs to act now.

4.12 THE HUMAN COST OF CLIMATE CHANGE

Global warming and climate change is already causing a great burden on civilization and a lot of human suffering. During 2020 almost 30 million people had to leave their homes because of weather-related disasters across 140 countries and territories.[115] Storms, floods, and cyclones/hurricanes/typhoons caused most of the damage. This is the highest figure recorded since 2012 and three times the number of displacements caused by conflict and violence. The highest number of disaster displacements took place in Bangladesh (>4 million), China (4 million), India (5 million), and the Philippines (>4 million).

As the IPCC report has found, extreme events, such as droughts, fires, and floods, are increasingly compounding each other, likely as a consequence of human influence. This is true also about human-made disasters. In a project at a UN University[116] ten different disasters from 2020/2021 were analysed. The researchers found that even though the tragedies occurred in vastly different locations and do not initially appear to have much in common, they are interconnected to each other. These causes include greenhouse gas pollution, poor disaster risk management, and economic analysis that ignores environmental costs and benefits. The events include flooding in Vietnam and freshwater fish extinction in China. The August 2020 Beirut explosion was also pointed out in the Prologue. The interconnectivity between the Arctic heatwave and the Texas cold spell in February 2021 is described in 4.5. In a state that is used to warm weather for the whole year around 4 million people had no electricity for 86 hours as the temperatures were below the freezing point. This led to the deaths of 210 people.

> In 2020 almost 30 million people had to leave their homes because of weather related disasters.
> This is 3 times the number of displacements caused by conflict and violence.

Some of the root causes of disasters are in fact influenced by the actions of people far away from where the event itself occurs. The global demand for meat is a major reason for the record rate of deforestation and wildfires in the Amazon. Farmland is needed to grow soy, which is used as animal fodder. Our actions have consequences, for all of us.

Climate change explicitly or implicitly affects key aspects of human health, including allergic and respiratory diseases.[117] Increasing CO_2 and global temperatures have been linked to longer duration of the pollen season, increased pollen concentration, and broader geographical distribution of pollens leading to increased duration and severity of seasonal allergic rhinitis and allergic asthma. The authors of the *Allergy* paper[117] argue that *'the greatest threat to global public health is the continued failure of world leaders to keep the global temperature rise below 1.5°C and to restore nature.'*

Since water is crucial for human survival, people simply try to move if there is a severe drought. The World Bank has analysed a huge dataset of internal migration, covering nearly 500 million people from 189 population censuses in 64 countries from 1960 to 2015.[118] The report has found that water deficits are linked to 10% of the rise in global migration. Climate change is amplifying the water crisis and it is predicted that exacerbating droughts will affect some 700 million people towards the end of this century. The poorest often lack the means to migrate. Residents of poor countries are four times less likely to move relative to residents of wealthier countries. The poor people are trapped. Agricultural workers are often driven to migrate to urban areas. This will increase the threat to urban water supplies, which are already subject to heat stress and water scarcity. Recent examples of major cities that are approaching 'day zero' are Cape Town, Chennai, and São Paolo. Of course, smaller places are also affected but do not get the same media attention.

4.13 ACTIONS NEEDED

Even if we manage to phase out fossil fuels, we will not be able to decrease the CO_2 concentration in the atmosphere for future generations. The world has to adapt to the 'new normal'. Or vulnerability must decrease, both in wealthy and in poor countries.

- Cities must be planned to be more resilient to extremes in temperature, water supply, floods and so on.
- Balancing water supply and water consumption is necessary.
- Building insulation and affordable ventilation can be developed.
- Agriculture should adapt to changing climate by reconsidering what is the best crop to grow, and so on.
- Urgently transfer from a linear to a circular economy.

doi: 10.2166/9781789062908_0057

Chapter 5
The water perspective

Margarita Huamani lives in Chaquicoccha, a village in the Ayacucho region on the slopes below the glaciers in tropical Peru. She says: 'It is warmer nowadays and the sunshine is burning. The rainy periods are uncertain, and it has become more difficult to grow the food. There is hardly any water.' Water scarcity is acute

in Chaquicoccha. The glaciers used to serve as water reservoirs for the village, but they have disappeared almost completely. Weather has become more extreme. It has become necessary to find new ways of farming in order to survive. Margarita's husband Eugenio Callahua adds: 'And I don't wish to live in the city. I was born to live in the countryside, to farm.'

Diakonia (www.diakonia.se) is supporting the village in a water project. The weather has become more unpredictable. To protect the crops from sudden cold weather Margarita and Eugenio have planted trees. The water project has provided ideas about how to save water. Sprinklers have been installed to help when the drought becomes serious. (*Photo: Laura Ardila*)

Water is essential for all life. There are no substitutes. Water is not renewable, so we must take care of the same amount of freshwater that was available for the dinosaurs. So, the water is reused. The problem is that a growing population, climate change, increasing standard of living, energy extraction and generation, food production, and industrialization put a lot of pressure on water resources. Pollution and contamination of available freshwater sources will further decrease available water. Too often water is considered to be ubiquitous and taken for granted and the water is not given its true value. Water is not just an environmental issue. It is a fundamental issue at the heart of justice, development, economics, and human rights.

Still water is a great killer. Flooding and contamination kill millions of people every year. Most often we can do something about this. Water has been and still is a source of conflict between people, between regions and nations. Water can be considered synonymous with human power and influence. Historically, the most powerful nations and kingdoms were established around freshwater sources – rivers or lakes. Civilizations have collapsed because of sustained droughts, exemplified by the Tang (907 AD) and Yuan dynasties in China, the Maya empire (900 AD) in Meso America, and the Khmer Empire in Cambodia that peaked in the 13th century.

In Figure 5.1, key interactions between water and other resources are illustrated. As described in Chapter 4, climate change and population increase have significant impacts on water resources and water quality. However, feedback mechanisms – as described in Chapter 2 – are always present. Water pollution and water scarcity have a significant impact on public health and population, most obvious in less privileged regions of the world. Water supply and use must be part of a circular economy. Water should never be 'wasted': it is a resource that can be recovered and reused.

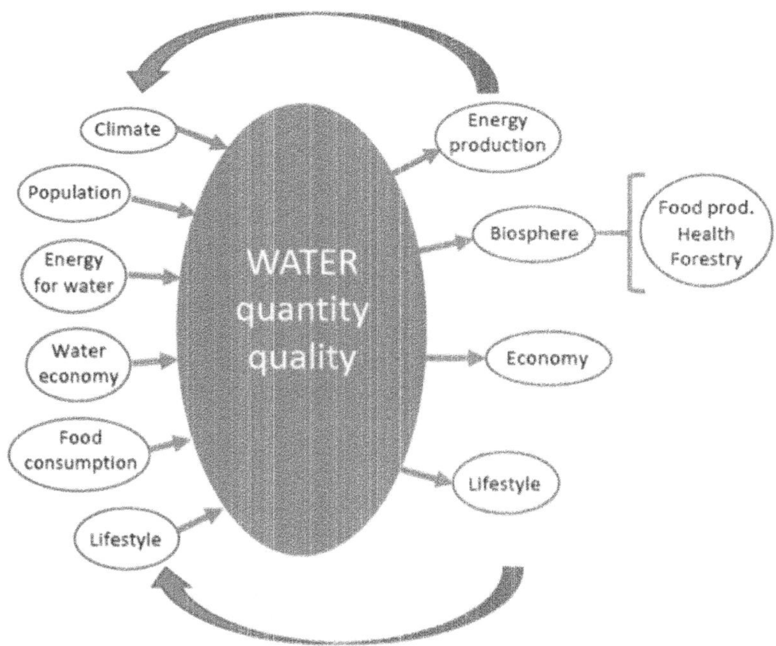

Figure 5.1 Interactions and feedback loops around water.

Obviously, water accessibility and quality depend on available energy sources. Energy, however, depends crucially on available water for fossil fuel extraction, for thermal power plants, and industrial process cooling. Energy production has impacts on water quality, so the feedback is obvious. In earlier work concerning water and energy[119], there are detailed descriptions of the water–energy interconnections.

Obviously, water is critical for the complete biosphere. Agriculture and food production is the main global consumer of water. Food and water are closely connected to human health, forestry, and nature. Agriculture production has a direct impact on water resources and water quality. What we eat has significant consequences for water consumption. Deforestation is a huge threat not only to biodiversity but to water resources as well. We must recognize that we are not the masters of nature but have to re-evaluate our dependence on nature. We are part of it and our attitude will influence water among so many other components. Slowly it is recognized that Nature has legal rights.

How water is valued is closely connected to water consumption. Too often the value of water has no relation to water tariffs. As a result, water is often over-consumed or wasted. Water resources are intimately connected to economy as well as to lifestyle. We need to re-evaluate our attitude to nature.

5.1 HOW WE GOT HERE

Worldwide, agriculture accounts for 70% of all water consumption, while 20% is used by industry and 10% in domestic life.

5.1.1 Drinking water

Some 2 billion people still lack access to safe drinking water. Despite this, Figure 5.2 illustrates that much progress has been made during the last 20 years. Around 62% of the global population had safe drinking water in 2000. By 2020 the number had increased to 74%. While the total population increased

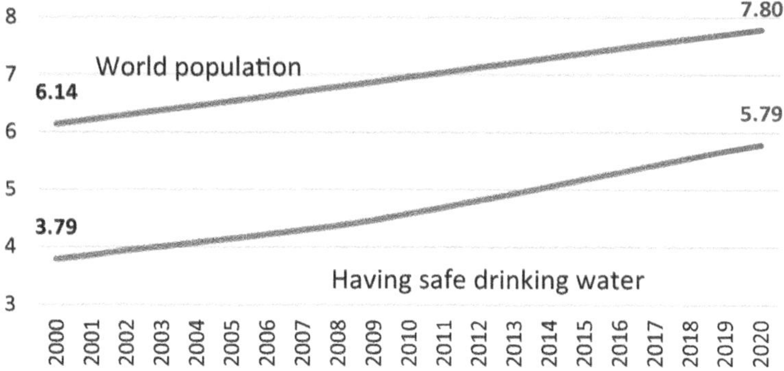

Figure 5.2 The world population (billions) having access to clean drinking water compared with the total world population since 2000[120].

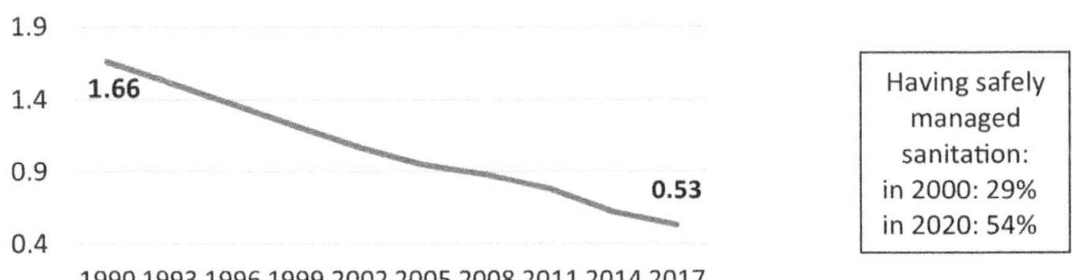

Figure 5.3 Number of deaths (millions) globally among children less than 5 years old, caused by unsafe water[121].

by 1.66 billion, the number of people having safe drinking water increased by 2.0 billion.

5.1.2 Sanitation

Too many people still lack safe sanitation, defined as improved facilities which are not shared with other households and where excreta are safely disposed *in situ* or transported and treated off-site. In 2000 only around 29% of the world population had access to safe sanitation; the percentage increased to almost 54% in 2020. In population numbers, this means that 2.45 billion people got improved sanitation, an increase from 1.76 to 4.20 billion or 2.4 times. Still, almost half the world population do not have access to safely managed sanitation. Among them, some 2 billion people have access to basic or limited sanitation that misses some criteria of safe sanitation. Unsafe sanitation is a huge health problem among the poorest people and is estimated to be the cause of around 775 000 deaths every year. Unsafe water still causes too many deaths among children, but the situation has improved during the recent decades, as shown in Figure 5.3.

5.1.3 Water stress

Over the past century global water use has increased twice as fast as the population, and it is still increasing in most sectors. Climate change will have a significant impact and will increase the risk of droughts. Since 1980 the number of people experiencing high water stress has doubled to over 2 billion people today[122], and the trend continues in the same direction. Global freshwater use during recent decades is illustrated in Figure 5.4.

5.2 THE STATE OF WATER RESOURCES

Water resources and quality are typically the primary indicators of climate change – like the canary in the mine. This in turn will affect many sectors, such as energy production, infrastructure, human health, agriculture, and ecosystems. The IPCC AR6 report emphasizes that the global water cycle will be intensified. The variability, the global monsoon precipitation, and the severity of wet and dry events will escalate. Much of the impact of climate

Having safe
drinking
water:
in 2000: 62%
in 2020: 74%

Global freshwater
use has increased
4 times since 1940

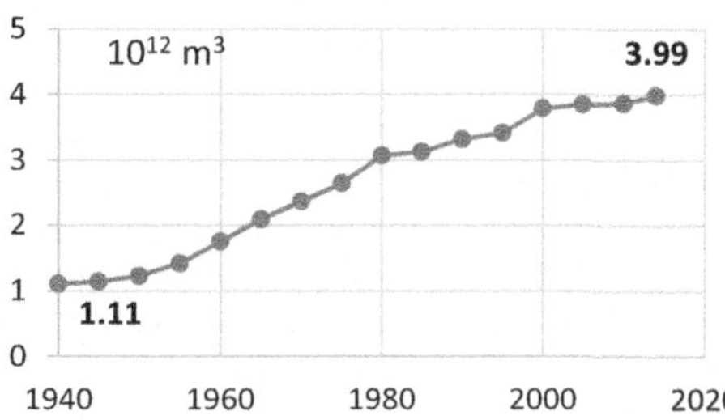

Figure 5.4 Global freshwater use development from 1940 to 2014, measured in trillion (10^{12}) m^3, is now 4 times higher, while the global population is 3.1 times higher, from 2.31 billion (10^9) in 1940 to 7.25 billion in 2014.

change on water resources will be manifested in the tropics, where most low-income countries are located, with potentially apocalyptic consequences for small island states, some of which could be wiped off the map. Mountainous areas are also exceptionally vulnerable through impacts on mountain glaciers and snow-caps.

5.2.1 Water quantity

Water is the top priority in adapting to climate change. In 2018 there were 2.3 billion people living in water-stressed countries and 3.6 billion people – 47% of the global population – had inadequate access to water during at least 1 month per year[123]. Other sources claim larger numbers, such as 4.0 billion people, or 52% of the global population[124]. Globally, a quarter of all cities are already water stressed and experience perennial water shortages[125]. Water scarcity tends to disproportionately affect the most vulnerable people and could displace an estimated 700 million people by 2030.

WMO predicts that the number of people having inadequate access to water will increase to more than 5 billion in 2050[126]. There is an increasing pressure on water resources. During the last 20 years stored terrestrial water has been lost at a rate of 1 cm per year. This includes all water on the land surface and in soil moisture, snow, and ice. This rate does not look high, but in a global perspective it is an enormous amount of water.

The gloomy development of water resources can be illustrated by the situation in Jordan. While the increase in demand in 2020–2021 compared to previous years is 70–80%, climate change has brought drier weather to the Middle East. Rainfall did not exceed 60% of the average during 2020–2021, according to the Jordan Water Ministry[127]. Jordan's population has doubled in the past 20 years to 10 million, including more than 1 million Syrian refugees. The water supply was 3400 m³/person/year in 2000. Now it is 80 m³. Consequently, only a fraction of the population has sufficient water supply.

As discussed in Chapter 4, climate change is likely to increase water demand while shrinking water supplies. Many ecosystems, particularly forests and wetlands, are also under threat, reducing biodiversity. Water supplies will be affected, not only for agriculture – which accounts for almost 70% of freshwater withdrawals – but also for industry, energy production, and even fisheries. Warmer temperatures increase the rate of evaporation of water into the atmosphere, in effect increasing the atmosphere's capacity to 'hold' water. As noted in section 4.4, the warmer, wetter air could also endanger human lives by blocking the cooling effects of our sweat. Increased evaporation may dry out some areas and fall as excess precipitation on other areas. The frequency of floods and droughts is increasing. In Europe, for example, the northern parts have been 10–40% wetter over the last century, whereas Southern Europe has become up to 20% drier. Over the last century annual river discharge has increased in Eastern Europe, while it has fallen in Southern Europe.

The seasonal variation in river-flow will change due to climate change. Higher temperatures will push the snow limit upwards in mountainous regions. Warming winter temperatures cause more precipitation to fall as rain rather than snow. Furthermore, rising temperatures cause snow to begin melting earlier in the year. This alters the timing of streamflow in rivers that have their sources in mountainous areas. With an earlier spring melt, there will be a shift in peak flow levels. As described in 3.2, due to declining glaciers there will be less water to compensate for low flow rates in summer.

Semi-arid and arid areas (such as the Middle East, Mediterranean, sub-Saharan Africa, and north-eastern Brazil) are particularly vulnerable to the impacts of climate change on water supply.

Low water resources and droughts have severe consequences on most sectors, particularly agriculture, forestry, energy, and drinking water provision. Activities that depend on high water abstraction and use, such as irrigated agriculture, hydropower generation, and use of cooling water, will be affected by changed flow regimes and reduced annual water availability. Moreover, wetlands and aquatic ecosystems will be threatened.

Water stress is one of the most serious current threats to sustainable development. The parameter *water stress level* is defined as the ratio between the withdrawal of freshwater from natural sources and freshwater renewal. Almost 30% of the global population – 2.3 billion people – are already living in countries that are water stressed. While the average global water stress is around 18% there are wide regional differences. Regions with the highest water stress in 2018 are shown in Figure 5.5.

At a country level, 35 countries are experiencing water stress of between 25–75% and 25 countries are considered seriously stressed, with figures above 75%. There are 16 countries with a water stress level above 100%, and the driest countries are Kuwait, Libya, Saudi Arabia, and the United Arab Emirates. Their demand for water is met mostly by desalination. However, there are more global water stress hotspots: Western North America, Western South America, the

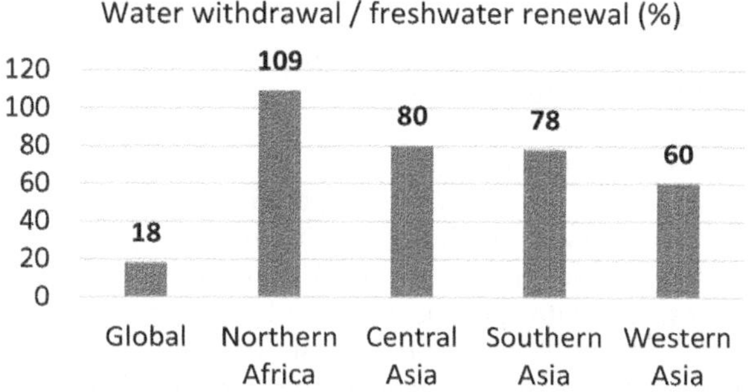

Figure 5.5 The regions with the highest water stress levels, expressed as the ratio (%) between water withdrawal and freshwater renewal.

Mediterranean area, the Sahel Region in Africa, Southern Africa, the Northern Horn of Africa, Central, South and Eastern Asia, and South-East Australia.

The 2018 edition of the United Nations World Water Development Report stated that nearly 6 billion peoples will suffer from clean water scarcity by 2050. This is the result of: (1) increasing demand for water, (2) reduction of water resources, and (3) increasing pollution of water, driven by dramatic population and economic growth. It is suggested that this number may be an underestimation, and scarcity of clean water by 2050 may be worse as the effects of the three drivers, just mentioned, of water scarcity, as well as of unequal growth, accessibility, and needs, are underrated.[128]

> UN: nearly 6 billion peoples will suffer from clean water scarcity by 2050.

As temperatures rise, people and animals need more water to maintain their health and thrive. Many important economic activities, like producing energy at power plants, raising livestock, and growing food crops, also require water. The amount of water available for these activities may be reduced as Earth warms and if competition for water resources increases.

5.2.2 Water quality

Water quality will be affected by increased water temperatures and a decrease in dissolved oxygen concentration. Also, water quality will suffer in areas experiencing increases in rainfall. Such events can cause problems in urban water systems, as sewer systems and water treatment plants are overwhelmed by the increased volumes of water. As a result, increasing runoff into rivers and lakes will wash sediment, nutrients, pollutants, trash, animal waste, and other materials into water supplies, making them unusable, unsafe, or in need of water treatment.

Water resources in coastal areas face risks from sea level rise, as saltwater can move into freshwater areas. As more water is withdrawn from rivers for human use, saltwater will move upstream. Drought as well as evaporation from hydropower reservoirs can cause coastal water resources to become more saline as freshwater supplies from rivers are reduced[129]. Rising sea levels and damaging impacts of storm surges will be an increasing threat to water infrastructure in coastal cities.

5.2.3 Natural disasters and water

Water availability is becoming less predictable in many places, and increased incidences of flooding threaten to destroy water points and sanitation facilities and contaminate water sources. The percentage of the global population at risk from

> 3 out of 4 natural disasters are water related

flooding has risen by almost a quarter since the year 2000. Flooding is the environmental disaster that impacts more people than any other, say researchers. Around 74% of natural disasters between 2001 and 2018 were water-related, including droughts and floods[130]. The frequency and intensity of such events are only expected to increase with climate change.

Water-related threats have been more frequent during the last 20 years. Flood-related disasters have increased by 134% compared to the period 1980–2000. Most of the flood-related deaths and economic losses took place in Asia. Warning systems were and remain inadequate in many places. The number and duration of droughts increased by 29% in the same period. Africa suffered most in terms of deaths[131].

The proportion of the population exposed to floods has grown by 20–24% globally since the year 2000, which is 10 times more than scientists previously thought. The reasons are both increased flooding and population migration. Nearly 90% of flood events occurred in South and Southeast Asia around the basins of major rivers including the Indus, Ganges–Brahmaputra, and the Mekong. The population inside flood-prone regions has increased to 58–86 million people[132]. The fact that people in many countries are moving into flood-prone areas rather than away from them has puzzled researchers. Between 2000 and 2015 the global population grew by over 18%. By comparison, in areas of observed flooding, the population increased by 34%. Worldwide, 2.23 million km^2 were flooded at some point between 2000 and 2018, affecting 255–290 million people. Climate change is changing the locations of flood plains to encompass more people. However, as the researchers emphasize, economics plays an important role. Places that have flooded tend to be cheap land for informal development, for example in northeastern India, and in Dhaka, Bangladesh. Naturally, the poorest people may not have another choice than to settle in a hazardous area.

There are many organizations[133], institutions, and researchers[134] concentrating on what can be done to build climate resilience from a water perspective and to adapt societies to withstand shocks and disasters. Since the theme of this book is to describe interactions, we will not describe in detail any mitigation and adaptation measures.

5.2.4 Glacier mass loss

In the period 2000–2019 the world's glaciers decreased their mass by 267 billion tons per year on average[135]. The dominating regions contributing to the

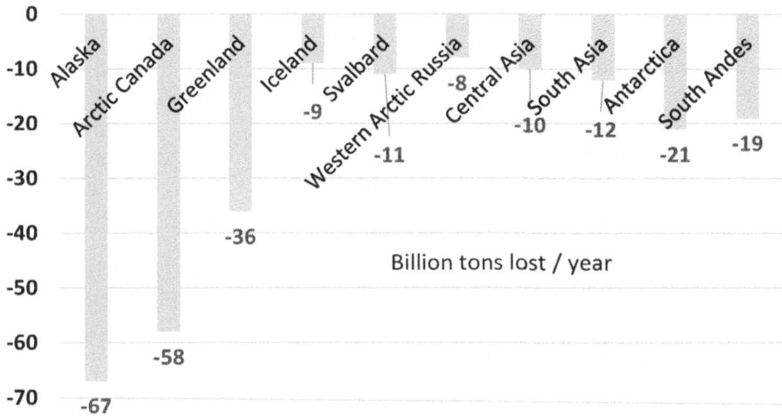

Figure 5.6 Glacier mass loss in the largest glacier regions of the world.

glacier mass lost are shown in Figure 5.6. Surprisingly, Alaska dominates the loss with around 25%. During the period there has been little precipitation and consequently the glaciers have not grown. This is a reminder that the glaciers around the world react differently to climate change. The regional differences depend on variations of temperature and precipitation between different decades and regions. Using satellite and aerial photos the elevation changes above sea level of more than 217 000 glaciers have been measured. The total area of these glaciers is some 706 000 km², equivalent to the size of Zambia, France, or Afghanistan. The glaciers play a significant role for both sea level rise and for the global water supply.

5.2.5 Water supply from glaciers
Obviously, the glaciers are most important water reservoirs. More than 1 billion people may be hit by water scarcity during the next 30 years, according to a special IPCC report[136]. Glaciers have an enormous importance for the global water cycle and are critical for the survival of millions of people. At the same time many glaciers are in regions that are inaccessible. There are some 200 000 glaciers in the world and only a few hundred of them have been studied at place. The IPCC special report finds that the glacier mass loss added around 20% of the sea level rise during the last decade. The glaciers contribute with around 10% of the rate of change of the sea level rise.

The Himalayan region glaciers did not lose so much depending on climate change but recently these glaciers are getting thinner. To illustrate the complexity: when temperature increases there will be more moisture in the atmosphere. This will lead to snow when the temperature is below 0°C. In Karakoram, in Pakistan, most precipitation is during the winter, so the glaciers may grow from snow and ice. However, in eastern Himalaya most of the precipitation is during spring and summer.

As Figure 5.6 shows, the Himalayan glaciers lose 22 billion tons of ice every year. This corresponds to 3.7% of the annual water consumption in China or 2.9% of the India water consumption[137]. The Himalaya is called the 'third pole', comparable to Arctic and Antarctic glaciers. These glaciers are the source of ten of the largest rivers in Asia, from Indus in the west of India and Pakistan to the Yellow River in China in the east. More than two billion people depend directly or indirectly on these glaciers, not only for drinking water supply, but also for hydropower, for providing food for all these people.

> More than two billion people depend directly or indirectly on water from glaciers

5.2.6 Sea level rise and land subsidence

According to IPCC recent rate of sea level rise has nearly tripled compared with 1901–1971. Thermal expansion is the dominating cause for the sea level rise, followed by the glacier melting. As a consequence of climate change there are around 200 million people living on land that is at risk of being flooded by rising sea levels towards the year 2100.

The problem of sea level rise is exacerbated by another phenomenon: land subsidence. Jakarta in Indonesia is an extreme example since the city is located at sea level. The sea level rise, combined with the subsiding of the land of up to 0.17 metres per year is increasingly serious. It is predicted that land sinks due to groundwater depletion, may threaten more than a billion people by 2040. Today about 40% of Jakarta already lies below sea level, according to World Bank estimates.

Land subsidence has often been misinterpreted by climate change sceptics, arguing that sinking land alone explains increased flooding in coastal areas. Sea level rise and subsidence are happening simultaneously. Sea level rise is a global issue caused by increasing ocean temperatures and melting of the world's glaciers, while land subsidence is a local problem. Land subsidence is a result of extracting something from underground. Groundwater extraction is the major cause, but the production of natural gas or oil will produce a similar effect. In many regions in the world, groundwater is being extracted faster than it can be recharged. In India 85% of drinking water comes from the ground, and in Europe 75% of the population gets drinking water from groundwater.

> Sea level rise and land subsidence are happening simultaneously.

Land subsidence occurs in several inland regions[138]. Mexico City's water supply depends heavily upon a local aquifer. The combination of the huge population of 20+ million people and inefficient water use, with more than 40% lost due to leakages means the aquifer is being emptied at an alarming rate. At the same time parts of the city are sinking some 30 cm per year. This also damages the water infrastructure, leading to more leaks and more water being withdrawn.

Tokyo is an example of good news. There the land began to sink more and more and reached a maximum in 1968 of 0.24 metres per year. Groundwater pumping had reached its peak at about the same time. In response, Tokyo's government passed laws limiting pumping. By the early 2000s the subsidence had decreased to only 0.01 metres per year. There is plenty of rain in Tokyo, but it is concentrated during just four months of the year. Therefore, at least 750 private and public buildings in Tokyo have rainwater collection and utilization systems. Tokyo is a role model for leakages. Investments in the pipeline infrastructure aims to reduce waste by leakage to only 3%[139]. Shanghai and other cities have tackled the problem by recharging their aquifers on top of limiting pumping.

Unlike Tokyo it may not be possible to change the main source of water. San Joaquin Valley in California relies on agriculture, and irrigation is crucial. Recent droughts have exacerbated the need for groundwater irrigation, and parts of the region have begun to sink by up to 0.60 metres per year. Water-intensive agriculture is part of the problem[140]. In 1939 John Steinbeck wrote his famous novel *Grapes of Wrath* describing the Oklahoma dust bowl (Steinbeck got the Nobel Prize in 1962). A poor family, the Joads, were driven from their Oklahoma home by drought and economic hardship. The farming family set out for California along with thousands of other 'Okies' seeking a better future. Is Steinbeck's story being played again in the USA or elsewhere?

Climate change has also increased extreme sea level events associated with some tropical cyclones, which have increased the intensity of other extreme events, such as flooding and associated impacts. This has augmented the vulnerability of low-lying megacities, deltas, coasts, and islands in many parts of the world.

5.2.7 Predicting future water resources

Our ability to predict climate impact on water resources depends on our understanding of the hydrological and ecological systems and how they are related to climate. First, there are great uncertainties in predicting climate changes. Typically, prediction is based on historical experiences. Our observations during a relatively short history may not be relevant for predictions. We can anticipate novel conditions which makes prediction even more difficult[141]. Therefore, it is increasingly important that we understand the mechanisms of the relations between climate, environment, and water.

5.2.8 Water conflicts

Since water resources are crucial for all aspects of life it is quite logical that conflicts will arise in the competition for water:

- *Biofuels*: IEA and other international organizations have highlighted the important role that biofuels must play in reaching net-zero emissions[142]. Biofuels' role in the global energy mix is still quite marginal. Traditional biofuel production has been fraught with challenges related to land use competition with food resources, water scarcity issues, environmental degradation, and adverse impacts on global markets.
- *Agriculture*: considering that irrigation uses around 70% of global water withdrawals, water scarcity can easily create tensions. A relatively new

phenomenon increasing the risk of conflicts is that food production depends on water availability. Water is probably the main reason for transnational large-scale land acquisitions (LSLAs), an ongoing agrarian transition from smallholder farming to large-scale commercial agriculture[143]. Most often the aim of LSLA seems to be to increase crop yields through the expansion of irrigation. Typically, investors prioritize land with access to surface and groundwater resources to support irrigation, while local users will be on the losing side.

- *Hydropower, national competition*: whenever a river crosses national borders there is an apparent risk for conflict about 'who owns the water'. There is a multitude of examples around the world[144], and we just mention a couple of them:
  - Ethiopia is completing the Grand Renaissance Dam in the Blue Nile. With a planned installed capacity exceeding 6 GW the dam will be the largest hydropower plant in Africa. Obviously, the Dam will influence water availability in both Sudan and Egypt downstream. Ethiopia is anxious to fill the Dam as quickly as possible. This is of course a great threat to Egypt since the Nile River water is a matter of life and death for Egypt.
  - The Mekong River (called the Lancang in China) in south-east Asia is shared by Laos, Cambodia, Vietnam, Thailand, and China. There is an enormous expansion of dam-building for hydropower, mostly in China and Laos. Countries downstream fear the negative impacts, like water shortages, flow alterations, sediment trapping, habitat destruction, and devastation of important agricultural areas and fisheries.
- *Hydropower* has, quite naturally, a great dependence on water. Its capacity is seriously impacted by global warming in already water scarce areas. Water reservoir dams will influence not only displacement of people but has public health impact, particularly in warm countries.

5.3 ENERGY FOR WATER SUPPLY AND WATER TREATMENT

Water supply, treatment, distribution, and reclamation require energy. The global average of the water-related electricity consumption by utilities is 4.0%, including water extraction and treatment, wastewater collection and treatment of the total electricity consumption[145]. This is most apparent in urban areas. However, there is a much bigger energy demand, for both electricity and natural gas, for heating and using water in homes and businesses. Also, in low-income regions there are energy demands, such as energy required for water trucking. Energy requirements for water supply and water treatment is discussed in detail in my previous book on water and energy[146]. Renewable energy offers great promises to provide water, in particular in regions outside the power grid as I described in an earlier book[147].

5.3.1 Source water abstraction and conveyance

In a recent literature review of energy demand for drinking water and wastewater systems in urban areas, 170 papers on energy demand in 34 different countries have been reviewed[148]. Supporting data are often not accessible or remain obscure. Data for the review are primarily concentrated in three countries: Australia, China, and the USA. Still, it offers some suggestions for energy use. The water-related energy consumption is found to be 1.7–2.7% of the total electrical energy[149].

If the end user energy for water is included, then the estimate is 12.6% of the total electrical energy in the USA[150]. Other publications[151] have found that energy related to water use within households, industry, and commerce accounts for 80–90% of the energy use for the urban water cycle. The energy use of water utilities for pumping and water treatment typically requires 9–10% of water-related energy. It has been demonstrated that 'water-related energy' is widely miscommunicated[152]. Utility energy use, while large, is often incorrectly over-emphasized. A range of other global, national, and regional documents have misinterpretations, where a statistic has been applied incorrectly or out of context. As emphasized by Kenway and colleagues[152], attributing all 'water-related electricity' to only the water utilities creates a risk of skewing policy to over-reliance on utility-focused solutions.

> Energy related to water *use* accounts for 80–90% of the energy demand in the urban water cycle

To estimate emissions from water related energy consumption is not straightforward. The electrical power can come from both fossil fuel power plants as well as hydro, nuclear, or renewable sources. Power for water heating comes from both natural gas and electricity.

Energy needed for water conveyance, from the source to the water treatment, can be widely different. When the source is close to the user locations, it is obvious that the energy demand is small. For example, in a relatively water-rich country like Sweden, the average pumping energy for water supply is 0.25 kWh/m³. In dry urban areas, however, there is a significant energy requirement. For example, in places like Southern California, in Johannesburg, South Africa, and in Adelaide, South Australia, water must be pumped long distances and high elevations to reach the final users. Energy requirements of 1.4–3.7 kWh/m³ for water conveyance have been reported.

5.3.2 Desalination

As an alternative to long water transports, seawater desalination is used in many areas. The energy issue is usually brought up as the obstacle for desalination. However, with energy requirement of 4–5 kWh/m³ for desalination, the energy cost is not significantly higher than the water transfer cost of 1.5–4 kWh/m³ in some regions. Furthermore, if renewable energy – solar photovoltaic (PV) or wind – can be used, then the environmental cost of desalination would be

> Daily global desalination production corresponds to around 12 litres/day/person

lower than long water transfers[153]. Today the global desalination production corresponds to around 12 litres per day per person of the global population of 7.8 billion people[154].

5.3.3 Irrigation and groundwater use

Irrigation is responsible for 70% of all freshwater withdrawals in the globe. The percentage may be even higher share for 'consumptive water use' due to the evapotranspiration of crops. We will describe the impact from irrigation from at least three aspects:

- The *water* perspective: risks for water conflicts and the consequence of energy use for water resources
- The *energy* perspective: competition between biofuel and food, and couplings to hydropower
- The *food* perspective: food production, and energy demand from irrigation.

Figure 5.7 presents a graphical indicator of content for the various irrigation perspectives.

Irrigation is used on agricultural land covering around 3 million km², corresponding to the size of Argentina or Kazakhstan. This is about 20% of cultivated land and 40% of global food production. There has been a ~70% increase in irrigated cropland area during the last 40–50 years. Consequently, water withdrawals for irrigation have increased from around 650 to 1400 km³/year (as a comparison, the total annual water use in Sweden, having 10 million people, is around 1 km³). This trend is explained largely by the rapid increase in irrigation development stimulated by food demand and by the continued growth of agriculture-based economies. It is estimated that without irrigation the global production of rice, cotton, citrus, and sugar cane would decrease by 31–39% and cereal production would decrease by 47%, representing a 20% loss of total cereal production worldwide[155].

> Water withdrawals for irrigation have more than doubled in 40–50 years

Food production has increased dramatically because of irrigation. However, it has also caused extensive environmental damage. With increasing water scarcity, there is an increasing competition for its use. This will increase the risk for food shortage in the near future.

> Food scarcity is often followed by water scarcity

Irrigation is a threat to water resources:

- *Groundwater*: excessive abstraction.
- *Inefficient* irrigation methods.

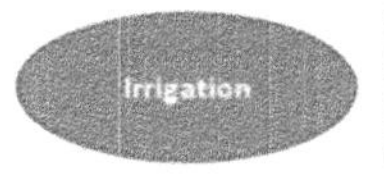

Figure 5.7 Irrigation interactions in various contexts described in various sections of the book.

- *Salinization*: this causes the worldwide loss of around 15 000 km^2 (corresponding to half the area of Belgium) of arable land per year and affects about 16% of all agricultural lands. In other words, an agricultural area of 1.3×1.3 km^2 is lost *every hour*, around the clock.
- *Saltwater incursion*: primarily in island or coastal communities.

In many regions the surface water is unusable for irrigation and farmers must use groundwater. However, surface sources of pollution can affect groundwater where the ground is permeable or where conduits to the water table are present. Sewage, landfills, and hazardous waste disposal sites are serious sources of groundwater pollution.

The world population is expected to exceed 10 billion people in 30 years. Obviously, agriculture must become more productive, resource efficient, and environmentally sustainable. More than 3 billion people live in agricultural areas with high levels of water stress or frequent droughts.

Pumping water more than recharge is increasingly a serious problem, especially in parts of China, India, and Bangladesh. It is also an increasing problem in high-income countries. For example, in some areas of the mid-West in the USA, the withdrawal from the High Plains aquifer is significantly times larger than the volume being recharged.

> Overuse of groundwater is not sustainable

Irrigation pumping uses 6% of global electricity. Experiences from India can illustrate development during the past few decades. As described in the prologue, millions of hand pumps were installed in India and other regions in the 1970s. Rapid applications of rural electrification, subsidized electricity, availability of cheap pumps, and local well-drilling have caused a tremendous increase of the number of tube wells for irrigation. The volume of groundwater abstracted in India increased from 10–20 km^3 before 1950 to 240–260 km^3 by 2000. All of this helped to lift millions out of poverty and hunger. Naturally, this consumption is not sustainable – and the link with energy is crucial.

To illustrate the crisis, groundwater tables in Gujarat, western India, have lowered by more than 170 metres during the last few decades, and are dropping at a rate of 6 m/year. Soon the groundwater will be beyond economic reach for most farmers. There is a similar dilemma in parts of China. When subsidies are no longer politically sustainable, and when global electricity prices rise, farmers will receive a 'double hit'. Some irrigation and farming practices have to be changed. In 2018 Gujarat had already consumed 68% of its groundwater and no signs had appeared that consumption would decrease[156]. In 2017 Gujarat used 12.3 billion m^3 for irrigation, or almost 400 m^3/second. This corresponds to the dry weather flow rate of more than a hundred water resource recovery facilities in large cities. The Gujarat Water Resources Development Corporation predicts that with current consumption only 6.8 billion m^3 would be available to meet future irrigation needs. On top of this, several districts in Gujarat have groundwater contaminated with fluorides, arsenic, iron, nitrates, and chlorides[157]. The situation is similar in Rajasthan in the north-west of India.

The electrical energy required for pumping is dramatically illustrated by the situation in India, where nearly 18% of electricity generation capacity is used for agricultural water pumping[158]. India has around 26 million agriculture pumps, including at least 12 million grid-based electric pumps and 10 million diesel-operated irrigation pump sets[159]. Farmers pay only an estimated 13% of the true cost of electricity[160]. The national burden of electric power subsidies is becoming too heavy. The subsidies encourage inefficient water use and contribute to overdraft of groundwater. As water levels drop, more power is needed to pump the water, thus increasing the energy requirement of water extraction.

India has announced plans to deploy about 2 million off-grid solar-powered irrigation pumps. However, the deployment has been sluggish, according to official statistics from the Institute for Energy Economics and Financial Analysis (IEEFA). In 2020, only about 250 000 solar pumps had been installed, just one eighth of the goal[161]. The initial investment in a

> Cheap electric power →
> too much pumping →
> lower groundwater levels →
> more energy use →
> dangerous water scarcity

solar pumping system is mostly higher compared to normal water pumps[162]. To solve this financial problem for farmers, the government offers various subsidy schemes. However, it is recognized that solar-based pumping causes a new risk for water resources. Since the operational cost of solar photovoltaic pumps is negligible and the availability of energy is predictable, it could result in overdrawing of water. One way to combat that unintended consequence is for the farmers who accept the subsidies to purchase the solar water pumps must switch to drip irrigation.

5.3.4 Wastewater treatment – water resource recovery

Used water is not waste so the wastewater treatment nomenclature is rightfully changed to water resource recovery facilities (WRRFs). On a national level, the electrical energy consumption for water and WRRF operations is in general quite small. The annual specific power consumption of a state-of-the-art WRRF is of the order 20–45 kWh/PE (population equivalent or unit per capita loading). The lower

> Water resource
> recovery typically
> requires <5 W/
> person

figure applies to larger plants, since the specific power consumption is usually higher for smaller plants. This corresponds to a load of 2.3–5 W/person, which is of the order 1% of the average electricity consumption in high-income countries (see Figure 6.11). It may be argued that energy savings in the WRRF industry cannot solve the global energy crisis. This may be true, but the energy cost for water operations is an increasing and significant part of the operating costs.

Power consumption depends not only on a plant's size, but also on its design. Plants with nutrient removal, such as nitrogen and phosphorous removal, will consume more energy than plants with only organic carbon removal.

There are significant opportunities to produce energy from used water. The influent to a WRRF contains 2–4 times the amount of energy required for the treatment process. Most of this energy is present as organic carbon and as thermal power. By extracting heat content from the effluent water and using the organic carbon in anaerobic digestion, there is a potential to design a 'zero-power' WRRF. Anaerobic digestion of sludge containing organic substrates will produce biogas. Co-digestion of sludge with food waste is an option to increase the overall biogas output. Biogas consists mostly of methane (50–75%) and CO_2 (25–50%) and small amounts of ammonia nitrogen, hydrogen (H_2), and hydrogen sulfide (H_2S).

Recognizing that there are more than 2 billion people without access to safe sanitation, there are still viable decentralized solutions. Using anaerobic digesters to generate biogas from collected waste, and then consuming this biogas for household energy needs could reduce indoor air pollution. This is in line with the SDG 7.1.2, clean cooking for all. Using biogas instead of wood can prevent deforestation.

The operational energy efficiency can be significantly improved by advanced instrumentation and control. This is covered in detail elsewhere[163] and is outside the scope of this text.

As noted in 4.10, it is well-known that direct nitrous oxide emission occurs in WRRF operations. This looks like the dominating carbon footprint. There is also a risk of direct methane emissions from anaerobic digesters and anaerobic sludge storage tanks[164]. The largest electrical energy consumption in the plants is related to pumping and aeration. Naturally, renewable energy gets increasing attention globally to reduce emissions from fossil fuel electrical energy[165].

Water recycling and reuse aim to increase water productivity. Several cities in water-stressed areas treat wastewater for use in irrigation or other uses. A few cities, like Singapore and Windhoek, Namibia, have developed systems to recycle wastewater for potable use[166].

5.4 WATER IMPACT ON HYDROPOWER

The interdependence between water and hydropower sometimes becomes perplexing. Lake Baikal in Siberia is home to a quarter of the world's surface freshwater and at 1640 metres is the deepest lake in the world. The lake reached its lowest point in 60 years in 2015[167]. The water level of Baikal had been high in 2014 and people feared floods. The Irkutsk hydropower plant was built on the Angara River, the only outlet from Baikal, and the dam made the level of the Lake one metre higher. The Lake became a reservoir. To eliminate the risk of flooding the hydropower plant was allowed to increase their release of water. However, precipitation became lower than normal, so the water level dropped to more than 0.1 metres below the critical level set by the government to prevent damage to the ecosystem. Nevertheless, the government allowed the hydropower plant to continue generating power and further lower the level of the Lake. As a result, several villages on the lake shores lost the water in the wells connected to the Baikal water level.

In the USA, Lake Powell is the second-largest man-made reservoir, located on the Colorado River, upstream of Hoover Dam. The region has suffered a 22-year drought, the worst in more than 1200 years. Local

Global warming – a huge threat to hydropower

scientists say that 'drought' is no longer the appropriate word to describe the situation. They call it aridification. In the summer of 2021, the water level in the dam fell to its lowest level on record, 1084 metres (3554 feet) because of drought. This is only 19.5 metres above the minimum level to produce hydropower at the Glen Canyon Dam, supplying power to 5.8 million customers[168]. With the current trend there is 33% probability to reach the minimum level in 2023. Further down along Colorado River there is a 22% probability that the 2025 water level in Lake Mead will drop to the level where no hydropower can be generated at the Hoover Dam. In response to the crisis in the Colorado River, the government in 2021 declared a water shortage in Lake Mead for the first time since Hoover Dam was completed in 1935. Currently the level is around 60 metres below its typical level. This will primarily affect the water supply to Arizona farmers. They use 74% of the water but represent only 1% of Arizona's economy. There is a significant risk that 40 million people from California to Wyoming to Mexico, depending on water from the Colorado River, will experience severe cuts of their water supply.

The water scarcity may create new conflicts around water. Utah, Wyoming, Colorado, and New Mexico are located upstream and are planning for new dams and pipelines from Colorado River. Under the 1922 Colorado Compact, they were granted these rights, but a hundred years ago the river flow was quite different from today.

Hydropower losses due to droughts have been reported from many other places than the Colorado River. Between 2020 and 2021 Washington State lost 11% of its generation capacity, while Oregon lost 16%, California 38%, and Alabama 16%[169]. When less hydropower is produced, then other sources need to compensate, including fossil fuel, nuclear, and renewables. The possible feedback to climate is obvious.

5.5 ECONOMICS, TARIFFS, AND THE VALUE OF WATER

Water pricing seldom reflects the real value of extracting water from nature and returning it without harming nature. Water is often extracted from aquifers having fossil water that is not renewed for generations. This takes place both in high-income and in low-income countries, for example in northern China (Beijing), India (Gujarat, Chennai), the West Bank and the Gaza strip, South Africa (Cape Town), and south-western USA. Logically it should be more expensive to extract a non-renewable water source than a renewable one.

Water subsidies may be motivated, but often they get unwanted consequences. Globally, 6% of water and sanitation subsidies go to the poorest 20%, while 56% goes to the wealthiest 20%[170]. Such subsidies will amplify inequality and reward inefficiency. Making water too cheap promotes overuse, which threatens the water service sustainability.

If local water resources are insufficient, one alternative is to pump the water from distant sources, already practiced in many regions, for example to supply water to Southern California, South Australia (Adelaide), South Africa (Johannesburg), or China (Kunming in the Yunnan province). Huge water transfers from the Yangtze River toward northern China is an example of massive water transportation. Not only the cost of transporting the water, but the environmental cost is substantial, both in terms of building the pipelines and energy use for their operation. If there were a carbon emission tax this cost would be significant and would influence the water tariffs.

Access to clean water is recognized as a human right. This immediately leads to the question: how 'much' is a human right? It is not a human right to waste water. Clean water requires treatment and a vast infrastructure and distribution do not come cheap. In many countries, it is undertaken as a public utility fully paid for by the taxpayer. However, the cost of accommodating growing demands is too high for the public purse. The affordability of water has become an urgent issue in many places, including in high-income countries[171].

> Water is a human right.
> Wasting water is *not* a human right.

If water is priced correctly according to market evaluations, there will be some who will have to use water so frugally it could pose a danger to personal and public health. In poor countries, even in the middle of cities, people are resorting to untreated groundwater for the daily needs. In these circumstances, outbreaks of cholera and other water-borne diseases are a regular occurrence, especially where a proper sewerage system is lacking. For public health and other reasons, every household should have access to treated, piped water – but at a price structure that incentivizes conservation.

In some cases, water is a common resource. A local community manages and uses water together. Different members may hold some rights. However, when the rules or community control are weak, 'open access' can apply. Some members of the community apply 'freeriding', which can lead to over-exploitation of resources – 'the tragedy of the commons'[172].

In many countries, both low-income and high-income regions, water is massively subsidized, subjecting it to serious undervaluing and severe misuse by individuals and industry. Yet, there is a need to ensure that every human has the right to get clean water. Water pricing needs to be revised in many places and countries. The closer the price of water approaches full cost of service; the better water can be valued.

Many places charge for water so that even the poorest people can afford a minimum amount of water use, the most valuable water that life depends on. Affordability is a major issue, and many cities and regions have resorted to utility credit programs or other means in their tariff structure to achieve affordability goals. Anything above this level should be priced according to the real costs it takes to make the water drinkable. To water a lawn in a water-scarce area is not a human right and should be charged accordingly. Simply

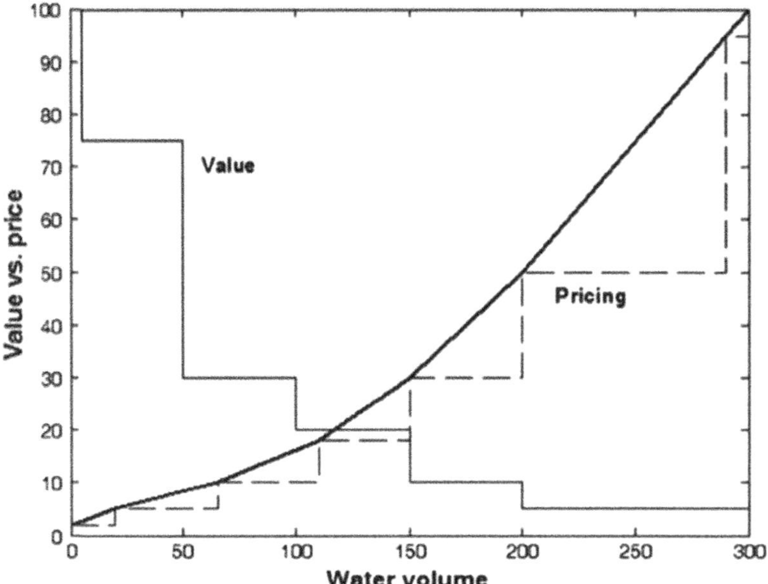

Figure 5.8 The perceived value of water compared to a structure of the tariffs[173].

expressed, we should pay less for the necessary water need and more for the 'luxury' needs. This is qualitatively illustrated in Figure 5.8.

It is crucial to find some technique to determine the economic value of water services. Of course, this is related to the willingness to pay. It looks as if too many countries and governments consider water as a limitless natural resource that can be freely exploited and used by any authority or by the landowner. Unlike any other commodity – such as oil – there is hardly any market defined for water. Only the cost of pumping, treating, and distributing the water is commonly charged. There is mostly no cost specified for the degradation of the water ecosystem: to pump water from a river or other surface water makes no difference than pumping fossil water from an aquifer. The water seems to belong to everyone, and nobody has the responsibility.

Some countries have recognized that the tariff should encourage efficient use of the water and have a tariff structure motivated by Figure 5.8. This is the practice in Greece, China, India, and in many African countries. Figure 5.9 shows progressive tariffs in some Asian and African cities. The tariffs are continuously updated, but the principle remains.

The price structure can be illustrated from Durban, South Africa. The first 6 m^3/month (200 litres/day) per household of water are free of charge for property chargeable values of less than Rand 250 000 (US\$ 14 500), a lifeline to many. For more expensive properties, the tariff is 21 Rand/m^3 (US\$ 1.2/m^3) from the start. There is no fixed charge. For a consumption higher than 6 m^3/month, the tariff is the same for everybody. If the water use is more than 30 m^3/

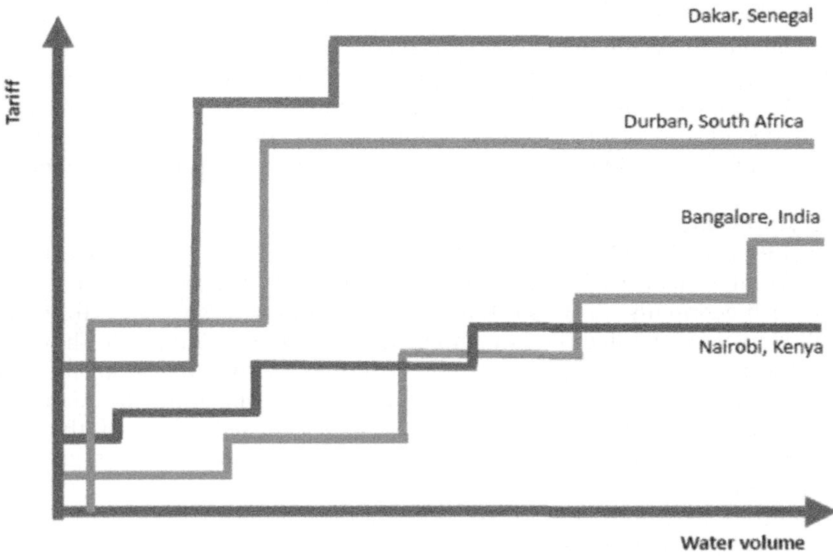

Figure 5.9 Progressive tariffs in some cities in Africa and Asia[174].

month, any household must pay around 52 Rand/m³ (US$ 3.0/m³). Obviously, the purpose is creating disincentives for overuse. So, the utility must balance the right for everybody to get clean water and to collect revenues to cover costs. Subsidizing water strategies depend a lot on the relation between connected low-income and high-income households. Cross-subsidies from high-consumption (high-income) to low-consumption (low-income) households are effective only if enough customers use the higher blocks[175].

The other extreme structure of water tariffs is applied in many places, where the fixed connection cost may be as high as 90% of the total water cost. Naturally, this tariff structure does not provide an economic incentive for the user to save water.

Water tariffs should reflect not only production costs but also water and energy availability. Conventional economics does not calculate the cost for water degradation. The economics of water sometimes resembles the economics of ecology. Biodiversity brings stability to ecosystems, which provide a wide range of 'services' that businesses rely on yet received free of charge. Because there is no financial cost for these services, they have been treated as being without value. This has resulted in corporate decisions that damage the ecosystem, reduce biodiversity, and leaves fewer degrees of freedom for future action. The solution is to value these ecosystem services so that they can become part of planning and decision making. This has nothing to do with corporate social responsibility or green agenda, it is hard-nosed economics. If the economic values of those services are considered, decisions automatically promote sustainability.

5.6 FOOD CONSUMPTION IMPACT ON WATER

The demand for water in agriculture has been commented in 5.2 and 5.3. Another type of information is how agricultural water withdrawals as a share of total water withdrawals relate to GDP per capita[176]. Generally, withdrawals tend to decrease at higher incomes. In low-income countries agriculture forms a higher share of total GDP and a larger share of agricultural employment.

5.6.1 Water use

It is a well-established fact that water requirements are significantly different for different food types. The concept of virtual water was invented in 1993 by the British economist Professor John A. Allan from King's College in London (in 2008 he was awarded the Stockholm Water Prize)[177]. Figure 5.10 shows the virtual water footprint for common food products. This water footprint is the sum of water requirements across the full value chain, including the quantity of water polluted in the agricultural production. For meat production, the water requirement includes both the water for the animal and for the crops grown as animal feed.

Countries can both import and export virtual water through their international trade relations. A water-rich region like Europe is a net importer of virtual water from regions with severe water scarcity. Also North Africa, the Middle East, and Japan are net importers, while North and South America, Southern Asia, and Australia are net exporters of virtual water. Regions leading in beef and grain exports are the top exporters of virtual water. The average total consumption of virtual water is discussed in the next section.

> Europe, a water-rich region, is a net importer of virtual water from regions with severe water scarcity.

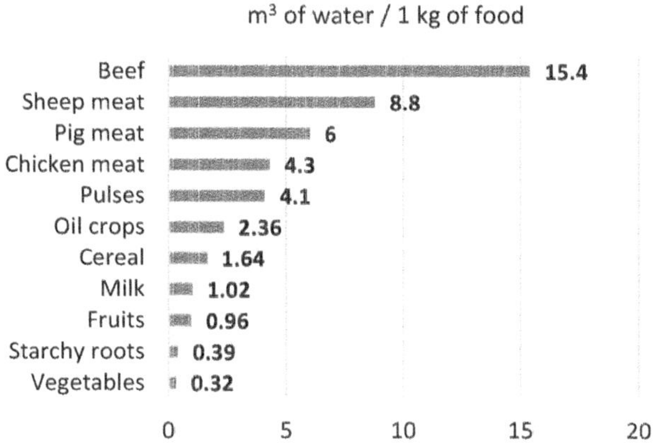

Figure 5.10 Water requirement (m³) to produce 1 kg of food[178].

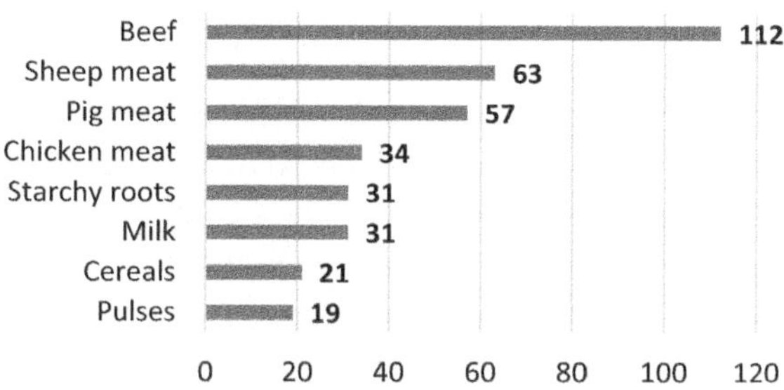

Figure 5.11 Water requirement (litres) to produce 1 gram of protein.

Animal products not only require more water per kg. A meat diet also requires significantly more water to deliver a certain amount of nutrients, as shown in Figure 5.11.

Meat is not only an 'inefficient' food source. It requires more energy, water, and land to produce than any other food source. It can also be demonstrated by its *protein efficiency*. An efficiency of 50% means that 50% of the protein in the animal feed is effectively converted to animal protein. The rest is lost during conversion. The meat efficiencies are:

> Meat is an 'inefficient' nutrient source.

- for beef: 3.8%
- for pork: 8.5%
- for poultry: 19.6%.

Expressed differently: we must feed cattle 25 kg of feed to produce 1 kg of beef.

Since meat consumption is increasing rapidly in high-income regions, the impact on water consumption is significant. Globally, every *hour* around 250 million m³ (=0.25 km³) of water is used for meat production (compare Section 4.12). This is the still incomprehensible flow of 70 000 m³/second or 28 full Olympic Swimming pools every second, around the clock.

5.7 LIFESTYLE AND WATER CONSUMPTION

Agriculture water use is crucial both to future food production and to water resources in general. According to FAO, the world needs to produce 60% more food by 2050, while conserving natural resources. Industrial and domestic use of water must compete with the food production. Obviously, lifestyle is a critical factor, in terms of water use at home, of food habits, and of consumption patterns.

Our *water footprint,* as discussed in 5.6, includes not only tap water and virtual water used to produce food, but also electricity, gasoline/diesel, as well as home goods. There are ways to estimate the water

> Water footprint is closely related to lifestyle.

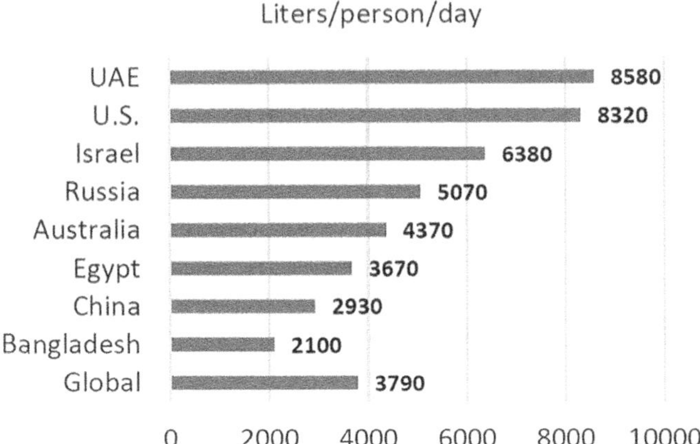

Figure 5.12 Total water footprint for some countries[179].

footprint as in an online calculator www.watercalculator.org created by Grace Communications Foundation. The previous section emphasized virtual water in our food but there are many other points to consider.

- Driving a gasoline or diesel car adds virtual water, of the order 200 litres for every 100 km.
- Producing clothes requires plenty of water. For example, it takes an estimated 2000–8000 litres of water to make one new pair of jeans, because of the water it takes to grow the cotton.
- There's also water used to produce our electronics, our televisions, and our phones.

Primarily, we can influence the water footprint by being aware of it. Eating less meat, driving less, and shopping less will lower our water footprint as well as our carbon footprint and expenses. The water footprints vary between countries because they're based on consumption habits of people within each country, as well as that country's climate and water-use practices, as shown in Figure 5.12.

5.8 OUR ATTITUDE TO NATURE

Does Nature have rights? For a long time in our history, humans have been perceiving nature as something that needs to be conquered and used to satisfy the needs of humans. We looked at environment as an inexhaustible source of natural wealth that should be extracted for short term gain. Humans have been seen as a force that could dominate nature. The world has been perceived within a *paradigm of a machine* that serves the needs of humans[180]. Nature and all the parts of it are not 'things' or property but living beings with fundamental value and an inherent right to exist. The idea is centuries old and has been an inborn part of indigenous and other land-based cultures.

We have come to a point in history where we more than ever need to look at nature not as something that needs to be conquered and exploited but as a partner in life. If we understand that we are a part of nature, it will become clearer that a battle with nature is a war against humanity. E. F. Schumacher puts it like this in *Small is Beautiful*: '*Modern man does not experience himself as a part of nature but as an outside force destined to dominate and conquer it. He even talks of a battle with nature, forgetting that, if he won the battle, he would find himself on the losing side.*' We must attend to the living now and focus on what creates more life – in ourselves, among our fellow human beings and among all the living creatures on Earth.

The obvious fact that water is life should have a profound impact on how we extract, treat, use, and return water to nature. If water is in abundance, we seldom pay attention to any imbalances in nature. In our high-technology and urbanized lives, we often forget our connection to and dependence on nature. With increasing population and urbanization, higher demand from industries and food production, intensified by climate change and more erratic precipitation patterns, the balance between human water use and nature is threatened. '*There needs to be some kind of water stewardship that ensures that the urban and the natural water cycles work together seamlessly and without destroying values in either place.*'[181]. We need to acquire a better understanding of the relationship between urban water systems and the surrounding nature, not only in general or in global terms but in each specific location.

Pernille Ingildsen expressed it like this[182]: '*Something more than the "ultimate good control of water" should be defined: looking for something called "water stewardship". "Smart" or "intelligence" must be supplemented by something of a different dimension, such as "the best of humanity", "a caring respect for nature" and "poetic beauty".*' The following explanation of wisdom should define the ultimate goal of any individual or organization (*ibid.*):

'*To me, the difference between knowledge and wisdom is that wisdom is eternal and shows us a way of being in life – while knowledge is finite, being changed all the time. Knowledge has to be rewritten all the time – while wisdom is our sounding board for securing that our decision is based on a natural ground. Knowledge without wisdom is like water in the sand. Going beyond the famous Einstein quote; that you cannot change anything by using the same way of thinking, will demand of us to include our eternal wisdom coming from the heart.*'

It all boils down to a more beautiful, meaningful, and comprehensive understanding of our role and our common work to add value to the overall water cycle. Nature is part of the water cycle and this fact can have a profound impact on water operations. The legal movement for the rights of nature was proposed in a legal study, *Should Trees Have Standing?* by the law professor Christopher D. Stone. He was 'quite

How to acquire a caring respect for Nature?
Can Nature trust us?

seriously proposing' giving legal rights to nature.[183] Today, New Zealand's Whanganui River is a person under domestic law, and India's Ganges River was recently granted human rights. Bolivia and Ecuador have passed laws granting all nature equal rights with humans. In Ecuador, since 2008, the Constitution enshrines nature's 'right to integral respect' or Pachamama. This means that 'Nature has the right to exist, persist, maintain and regenerate its vital cycles'. In practice, that means that all persons, communities, peoples, and nations can demand that Ecuadorian authorities enforce the rights of nature. One of those rights is the right to be restored.

An historical victory for nature took place in the constitutional court in Ecuador on 2 December, 2021[184]. Los Cedros, a 'cloud forest' in the Andes in the north-western part of Ecuador, is one of the most biodiverse parts of the planet, home to flora and fauna found nowhere else on Earth. Los Cedros was declared a nature reserve in 1988. Despite this protection the Ecuadorian state mining company, in cooperation with the Canadian Cornerstone capital resources, achieved the right in 2017 to explore within the rainforest. Now the law protecting Nature means that biodiversity is valued higher than possible economic profit from mining operations. The judge wrote that *'the risk in this case does not necessarily concern people .... but the eradication of species, destruction of eco-systems and a permanent change of natural cycles.'*

The right of nature will cause a conflict in western legal systems, where individual liberty and personal property are considered fundamental rights. If nature has rights, then humans have new responsibilities and must restrict some demands that unjustifiably impair the rights of nature. It boils down to: 'can Nature trust us?'

5.9 ACTIONS NEEDED

The global water situation is grave. Since water is life there must be a revolutionary change in the way we use water resources and how we pollute water. Without affordable and available clean water hardly any of the sustainable development goals can be reached.

- Clean water for everybody goes without saying. The payback is huge, in terms of less vulnerability, public health, child mortality, and other essential parts of life.
- Water is a human right, but it is not a human right to waste water. This should be reflected in the structure of water tariffs.
- Water and food are intimately connected. Our eating habits must change. Current meat consumption in the world is unsustainable, also from a water consumption point of view.
- Water is crucial for food production. Irrigation is the largest water consumer. There must be a revolution in irrigation practices.
- Groundwater cannot be consumed faster than it is renewed.
- Renewable electric power outside the grid is a key ingredient to provide water supply and water treatment. It can be made affordable even for the poorest.

- Glacier losses will create huge water supply problems for millions. There are ways to decrease the agriculture vulnerability. This requires economic support from wealthier nations, education for farmers, and new considerations of what can be grown.
- Industrial and municipal pollution is devastating in many countries. Treatment of polluted water should have a high priority.
- Polluted water is not waste – it is a resource that should be recovered.
- Our lifestyle based on consumerism is devastating for the world water resources. Water, economy, and lifestyle are intimately connected.
- Can Nature trust us?

doi: 10.2166/9781789062908_0085

Chapter 6
The energy perspective

Nenibarini Zabbey, professor of hydrobiology, University of Port Harcourt, Nigeria, is a genuine water hero. For more than a decade Zabbey has been in the frontline struggling for environmental justice in the Niger Delta. It is a true 'David and Goliath' story, where dedicated individuals strive to hold powerful oil companies accountable for reckless management of the environmental impact of their operations. Zabbey has tirelessly monitored and documented water quality to provide useful insights and detect changes in the Niger Delta. He has made truly personal sacrifices in this ongoing battle, living with various threats because of his struggle for a better future for his people. His pre-spill research datasets became the basis of determining the ecological damage in the Bodo Creek following two major oil spills in 2008 and 2009, that have now been followed by the world's largest mangrove clean-up and remediation project. The oil spill case was litigated in UK in 2011 and lasted for 4 years. In a 2020 report, Amnesty International stressed Zabbey's key role in the Bodo oil spill justice struggle[185]. He is also personally involved in the mangrove remediation project with unpaid voluntary ecological and management support.

Here is a scientist who daily, on top of all his academic duties, has held giant oil companies accountable for avoidable negative environmental footprints, risked his own life several times, not only while sampling water and sediment but also by the actions of powerful organizations that have been rightly questioned by their reckless activities. By coupling limnology with human rights, Professor Zabbey is more responsible than anyone for securing a world historical environmental justice victory of the highest magnitude with the unprecedented Bodo legal settlement.

Zabbey is the 2022 awardee of the ASLO (Association for the Sciences of Limnology and Oceanography) Ruth Patrick Award, honouring scientists who have applied the aquatic sciences towards solving critical environmental problems. (*Photo: private*)

Access to affordable and clean energy is fundamentally important for a decent life. The many interconnections to energy are described in Figure 6.1. It is evident that energy consumption has a fundamental impact on global warming. Also, climate change has an impact on energy availability and need. Energy extraction

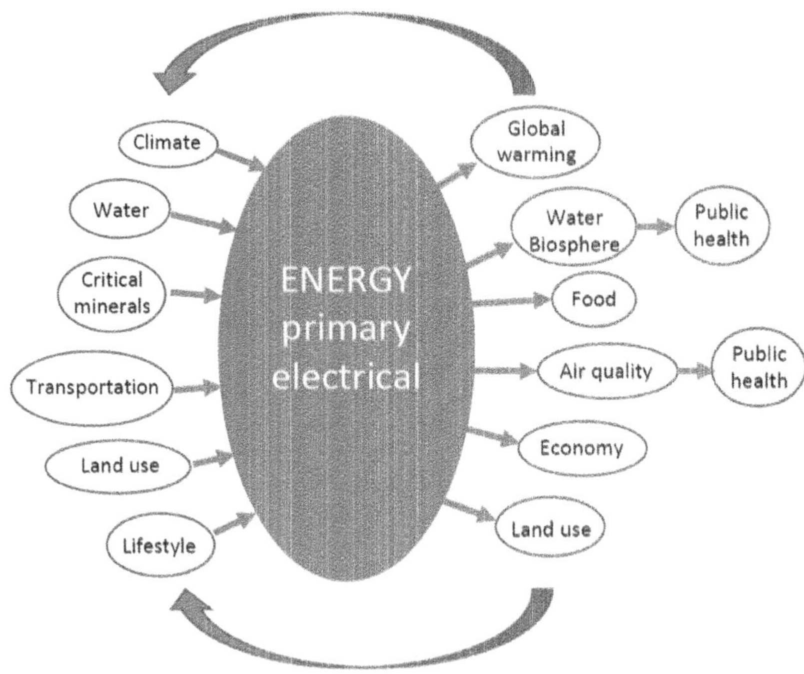

Figure 6.1 Interactions and feedback loops around energy.

and electrical energy generation depend crucially on water. Renewable and carbon-free energy depends also on critical metals and minerals. Our modes of transportation have a large impact on energy use. Energy extraction as well as electrical power production depend on land use. For wind power sea area use is also important. This most often creates various conflicts of interests. Lifestyle has a fundamental impact on energy use.

Energy extraction and generation has a huge impact on water quantity and quality as well as many aspects of the biosphere. Food production depends fundamentally on energy availability. Too seldom public health aspects are prioritized when energy plans are decided. Emissions and smoke from fossil fuel use are causing serious air quality problems which in turn will affect people's health. Obviously, economic considerations are essential in all energy planning and consumption. Fossil fuel extraction has a huge impact on land use, as described in this chapter.

Like the water perspective, the energy perspective contains a multitude of feedback loops, as illustrated in Figure 6.1. As we get into the details, we will describe some of them. Still, the way forward is obvious: increase investments in clean energy and find ways to cut energy use.

6.1 HOW WE GOT HERE

Energy development during my lifetime is summarized here.

6.1.1 Global energy

The global energy production has increased more than nine-fold during my lifetime as illustrated in Figure 6.2. Oil production has grown a factor of 20 and has been the dominating fossil fuel source since the 1960s. Coal was the leading fuel source in 1940. Since then it has grown by a factor of 4 and is now levelling off. Natural gas provided only 6% of fossil fuel use in 1940 but contributes now

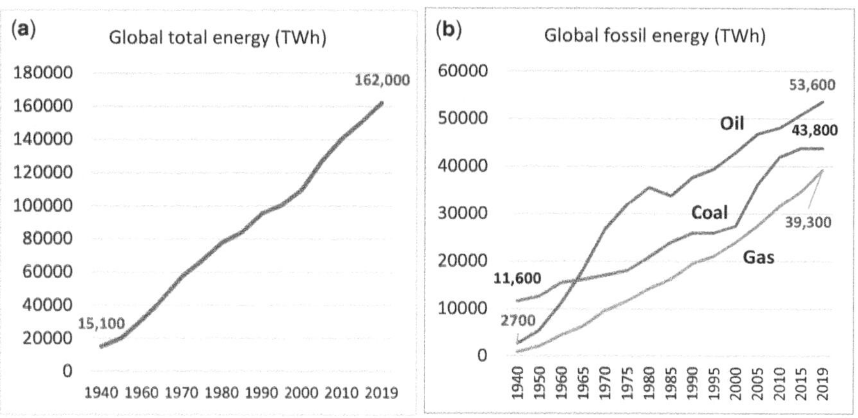

Figure 6.2 The global energy production (terawatt hours) based on fossil fuels from 1940 to 2019[187]: (a) total energy; (b) contributions of coal, oil, and gas.

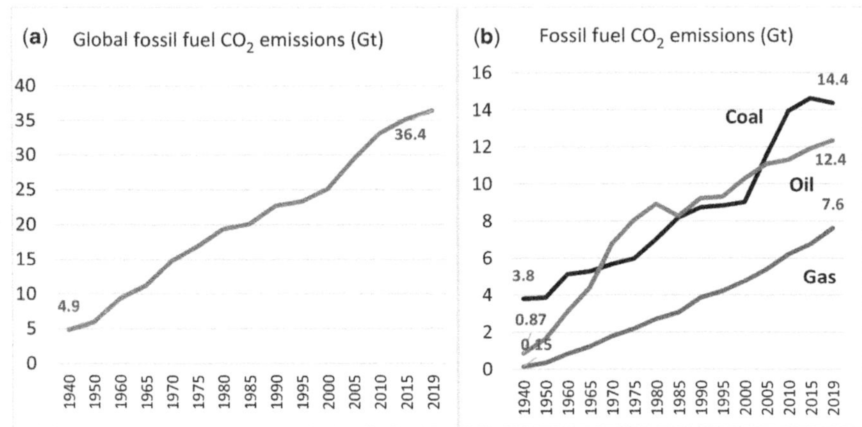

Figure 6.3 Emissions from burning fossil fuels from 1940 to 2019 (Gt)[189].

with almost 30% to the energy production, an increase of more than 45 times in my lifetime. The rate of change during the last 5 years is 13% for gas and around 5% for oil and coal. Fossil fuels contributed in 2020 with 83% of the total global energy use[186].

The fossil fuel consumption has caused an extraordinary increase of greenhouse gas emissions, by a factor of 7.5 times, during my lifetime, as illustrated by Figure 6.3. The most severe carbon footprint comes from coal, as shown in Figure 6.3b. According to the International Energy Agency[188] (IEA) the emission level for 2021 is expected to about the same as in 2019, after a decline in 2020 due to the pandemic. Coal, oil, and gas made up 94% of the total in 2019. Most of the rest is contributed by cement and flaring. The carbon footprint from coal is larger than from oil and gas, so its emission load is the highest even if the energy production is not the highest. The gas emissions have risen 50-fold from a low level in 1940. The oil emissions have increased a factor of 14 during the same period.

6.1.2 Electrical energy

The world is getting more and more dependent on electrical energy, as shown in Figure 6.4[190]. Global electrical energy demand has increased 2.5 times in 35 years. Coal is still the dominating fossil energy source (33%). Oil for electrical energy has been fairly constant over the 35 years, while the gas consumption has increased more than four-fold.

The global electricity mix is still dominated by fossil fuels, as shown in Figure 6.5. According to the IEA Electricity Market Report[191] for the first half of 2021, renewables had strong growth, some 8% in 2021. However, only about half the growth of global electricity demand 2021 can be satisfied by renewables. This results in a sharp increase of coal-powered energy, implying that there is a risk that the CO_2 emissions from electrical energy will reach record levels in 2022.

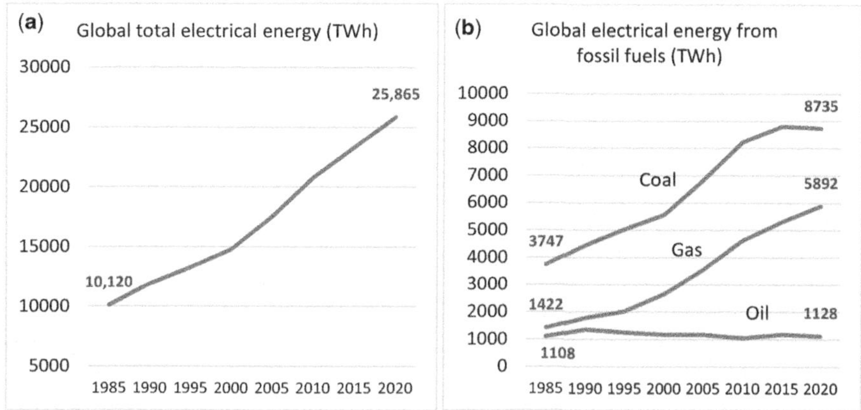

Figure 6.4 The development of global electrical energy since 1985.

Most of the demand comes from China and India. This raises a critical issue: is the increasing demand due to domestic needs or is it related to increasing consumption demand from high-income countries? The analysis of energy needs to include the whole supply chain, including lifestyle.

Figure 6.6 illustrates the development of non-fossil fuel electrical energy generation. Compared with 1940, hydropower generation has increased a factor of 23. Nuclear has levelled off since 2005. The Fukushima nuclear disaster in 2011 has had a significant impact on nuclear investments. Solar and wind are the dominating electrical energy investments today. Since 2015 solar energy has increased by a factor of 2.3 (232%). Wind had an earlier start and is still larger than solar. It increased by 92% since 2015 while other renewables increased 26%. It has taken decades of technology and commercial development of solar photovoltaic and wind before the rapid increases.

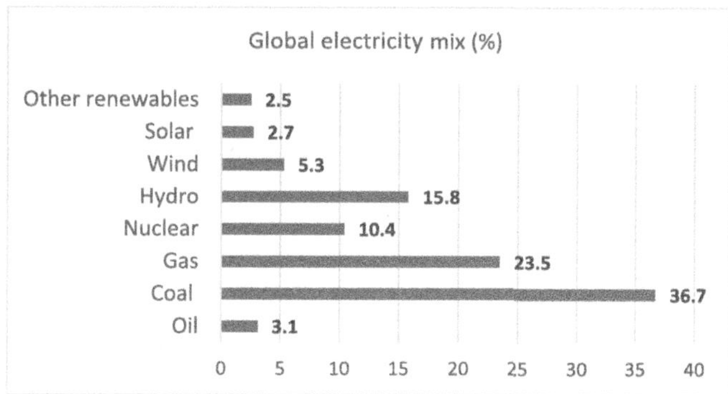

Figure 6.5 The global electrical energy mix in 2020. Still fossil fuels make up 63% of the electric power generation[192].

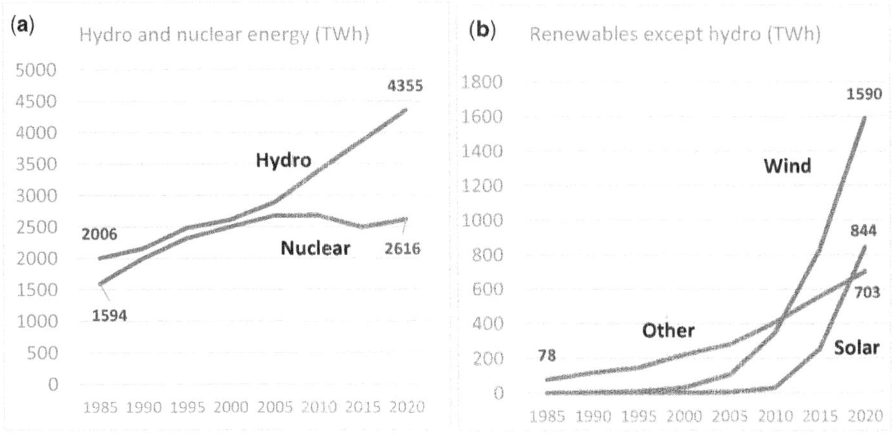

Figure 6.6 (a) Hydropower and nuclear power development during the last 35 years. (b) Development of renewable power.

Notably, in most countries no single key player has the long-term responsibility to ensure adequate production capacity. Who will build planned electric power production? Will import of electric power be possible when needed? Planning the energy transition is a crucial challenge.

6.1.3 Transport

The transport sector contributes about 20% of all CO_2 emissions[193], in other words some 8 Gt/year, as illustrated in Figure 6.7. Road transport accounts for 75% of the transport emissions, or 15% of total CO_2 emissions. In 2019 the airline industry contributed with 1.04 Gt, corresponding to 11.6% of transportation emissions or some 2.8% of total emissions.

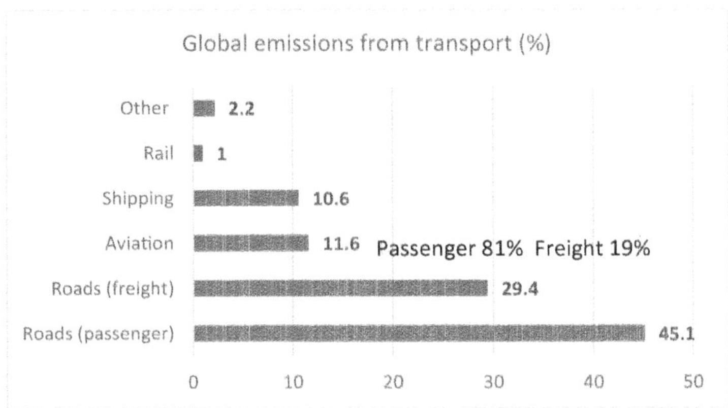

Figure 6.7 Global CO_2 emissions (2018) from transport[194]. The 2.2% 'other transport' includes moving materials like water, oil, and gas via pipelines. The numbers differ 1–2% depending on the data source.

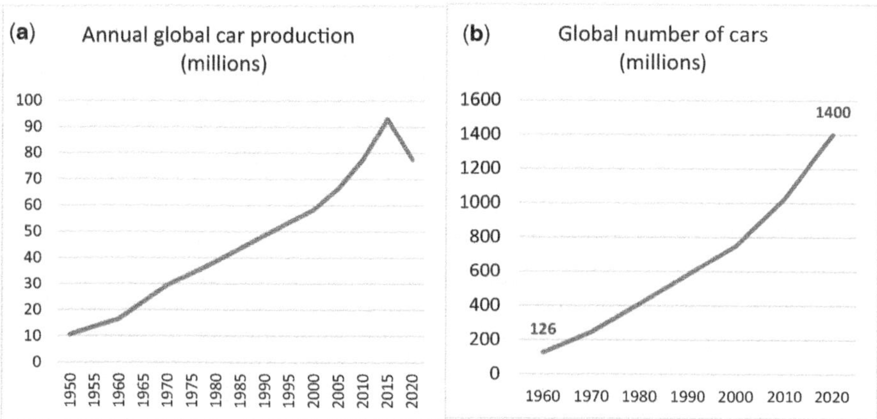

Figure 6.8 The global annual car production (including passenger cars, light commercial vehicles, minibuses, trucks, and buses) from 1950 to 2020[196]: (a) annual production rate and (b) accumulated number of cars.

Today there are around 1.45 billion cars in the world (Figure 6.8), and the number is increasing at a formidable rate with nearly 100 million cars produced every year. The production rate reached a peak value of 97.3 million cars in 2017 and has then declined back to the 2010 level. The dramatic drop from almost 92 million in 2019 to the 2020 level of around 78 million was to a large extent caused by the pandemic. The total number of cars exceeded 1000 million before 2010 and was 1450 million in 2021. Twenty-one per cent of the cars are in the U.S. Naturally all the road vehicles produce massive emissions (Figure 6.9). According to IEA, it will be a formidable task to reduce CO_2 emissions in the transport sector over the coming decades[195].

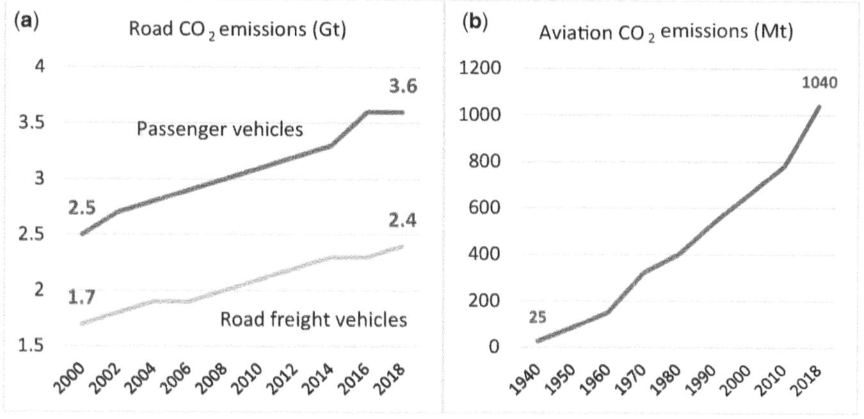

Figure 6.9 (a) Global emissions from passenger road vehicles (Gt) and from road freight vehicles after 2000[197]. (b) Global emissions from aviation from 1940 to 2019 (Mt)[198].

Global emissions from road vehicles after 2000 and from aviation from 1940 to 2019 are shown in Figure 6.9. The aviation numbers do not include non-CO_2 climate forcing or a multiplier for warming effects at high altitudes. Some 60% of the aviation emissions come from international flights and 40% from domestic flights. In the period 2000–2018 emissions from shipping increased from 0.6 to 0.9 Gt, while rail transportation has remained at the 0.1 Gt level.

6.2 ENERGY CONSUMPTION TODAY

Primary energy consumption comes from a mixture of coal, oil, gas, nuclear, hydropower, wind, solar, and other renewables. Primary energy supply is defined as energy production plus energy imports, minus energy exports, then plus or minus stock changes. Figure 6.10 displays the primary energy per capita and year in some high-income and some low-income countries in various parts of the world. There is a clear relationship between energy and GDP and the figure illustrates the clear indication of the different living conditions in various parts of the world.

The *electrical* energy consumption comparison between high-income and low-income countries, shown in Figure 6.11, visualizes even more than in Figure 6.10, the highly different living conditions in various parts of the world. As a comparison, the average consumption of one family's fridge in the USA is 450 kWh/year.

> In high-income countries we consume more than hundred times more electrical energy than in the least privileged countries.

Should we be surprised that people are trying to escape poverty from Central America to the USA or from Sub-Saharan Africa to Europe in their search for a better life? The unfair life in different parts of the world becomes even more absurd, considering that the poorest people pay the highest price for global warming.

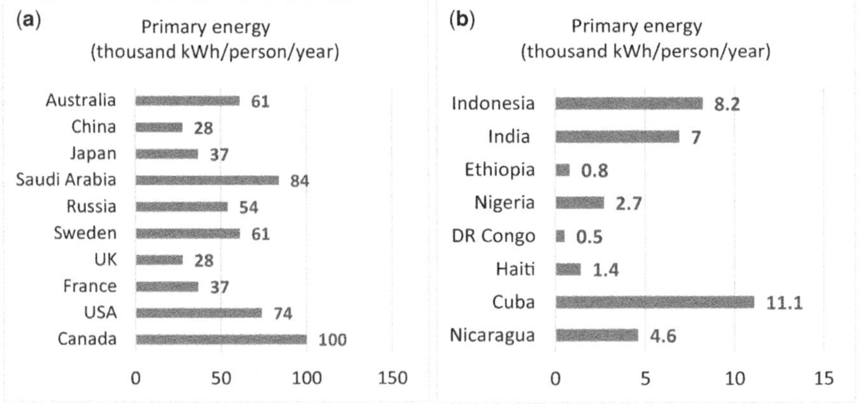

Figure 6.10 Primary energy consumption (thousand kWh/person/year) in some high-income countries (a) and some low-income countries (b) in various parts of the world[199].

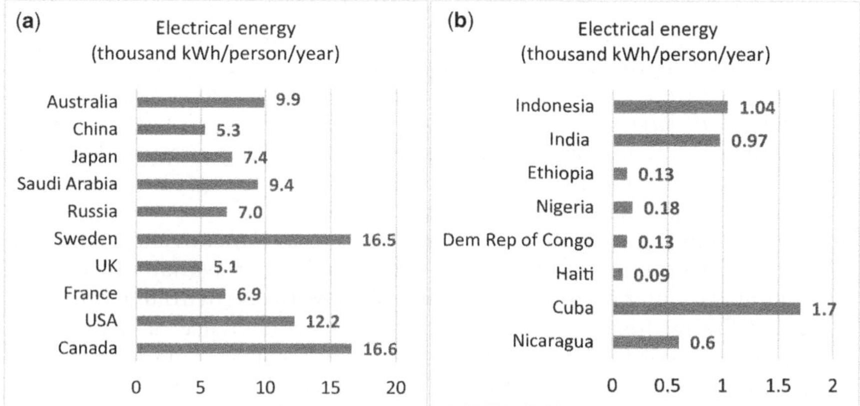

Figure 6.11 Electrical energy consumption (thousand kWh/person/year) in some high-income countries (a) and some low-income countries (b) in various parts of the world[200].

6.3 CLIMATE AND ENERGY INTERACTIONS

The apparent impact of energy from fossil fuels has been described. However, there is also a reverse impact. At least three major climate trends have energy production consequences:

- Increasing air and water temperatures,
- Increasing water scarcity in some regions and seasons,
- Increasing intensity and frequency of storm events, flooding, and sea level rise.

Increasing temperatures, decreasing water availability, more intense storm events, and sea level rise will each independently, and in some cases in combination, affect the ability to produce and transmit electricity from fossil, nuclear, and existing and emerging renewable energy sources. These changes are also projected to affect demand for energy and as well as the ability to access, produce, and distribute oil and natural gas.

- *Thermoelectric power generation requires water* . For many decades it was assumed that cooling water resources were available and reliable. Today there is a growing and recognized operational risk due to decreasing water availability and increasing ambient air and water temperatures. These reduce the efficiency of cooling, increase the likelihood of exceeding water thermal intake or effluent limits that protect local ecology, and increase the risk of partial or full shutdowns of generation facilities. Water temperatures in rivers and lakes are projected to increase with further increases in air temperature. Several power plants have already been forced to shut down in the USA, India, Brazil, France, and other countries due to lack of water or high-water temperatures compromising cooling processes. Thermal power plant projects are being re-examined to safe-guard their automated operation due to their impact on regional water resources and their vulnerability to climate impacts. This forms a major step in industrial digitalization. For example, during the 2003 summer heat wave in Europe, more than 30 nuclear power plant units in Europe were forced to shut down or reduce their power production[201].

- *Energy infrastructure located along the coast* is at risk from sea level rise, increasing intensity of storms, and higher storm surge and flooding, potentially disrupting oil and gas production, refining, and distribution, as well as electricity generation and distribution.
- *Oil and gas production*, including crude oil and unconventional oil and gas production, are vulnerable to decreasing water availability given the volumes of water required for enhanced oil recovery, hydraulic fracturing, and refining. Rises in temperature will either increase or decrease the access to fossil fuel resources, and there is an increasing temptation to consider supply of oil and natural gas from the Arctic regions.
- *Hydropower, bioenergy, and concentrating solar power* can be affected by changing precipitation patterns, increasing frequency and intensity of droughts, and increasing temperatures. More recurrent and longer droughts are threatening the hydropower capacity of many countries, as described in 5.4. Still another problem is extreme amounts of rain.
- *Electricity transmission and distribution* systems carry less current and operate less efficiently when ambient air temperatures are higher. They may face increasing risks of physical damage from more intense and frequent storm events or wildfires.
- *Wind and solar power* availability or productivity depends on changes weather patterns, cloud cover, wind resources, etc.
- *Weather-related disruptions*, such as changes in storm frequency/intensity, will influence energy infrastructure (for example, oil and gas drilling, pipelines and refineries, and power lines) and continuity of energy supply.

In many regions with colder climates today, there will be a reduced demand for heating and a strong increase in energy demand for cooling. The annual population-weighted number of heating degree days (HDD; see Glossary) in Europe has decreased around 6% between the periods 1950–1980 and 1981–2017 and the negative trend continues. The largest decrease has been in northern Europe. Similarly, the number of cooling degree days (CDDs) increased by 33% between the same periods and the increasing trend continues. The largest increase of CDDs occurred in southern Europe[202].

In Europe it is estimated that the decrease in HDDs will be much larger than the increase in CDDs in absolute terms. However, changes in CDDs have a larger economic and climate impact than changes in HDDs, since cooling is almost exclusively produced from electricity, while heating can be provided from energy carriers with lower cost and energy content (e.g., warm water). The general conclusions from Europe are also similar in other parts of the world.

> Global warming → more cooling, less heating → higher energy demand

Obviously, the energy performance standards of space cooling equipment need to be observed. All the technologies needed to address these challenges are available today. Appliances and cooling are the fastest growing uses of energy in buildings today, and their expansion is set to continue as incomes increase, more appliances are purchased, and millions of global households acquire air conditioners. To phase out inefficient equipment is naturally a matter of affordability and low-income countries would face great difficulties to achieve this. How to determine the setpoint of temperature will of course be a key factor in energy savings. The demand for space cooling is more than four times higher in advanced economies compared to low-income economies, even though emerging market and low-income economies have around four times more CDDs than advanced economies in a typical year[203].

Warmer winters in places where snow has been the norm will create another spectacular coupling between climate, water, and energy. According to the European Environment Agency, the length of snow seasons in the Northern Hemisphere has decreased around 5 days per decade since the 1970s. Normal snow is increasingly replaced with artificial snow in ski resorts. This is not only threatening the opportunities for anybody wishing to enjoy winter sports but also causing an increasing cost in terms of energy, water, and environmental burden. The requirement for artificial snow has come into focus as the Winter Olympic Games in Beijing were arranged for 2022[204]. In many places snow is also transported from other sites (by land or helicopter). For the Olympic Games in Sochi in 2014 snow stockpiling took place a full year ahead of competition.

Artificial snow is almost 30% ice and 70% air, compared to natural snow which is closer to 10% ice and 90% air thereby creating a grittier snowpack. To make artificial snow is a proven technology, but it requires water pumps, compressors, air plants, piping, snow guns, access to water, and electric power. Often, chemical or biological additives may be added to allow the water to freeze at higher temperatures. Salt and fertilizers are sometimes added to make the snow harder on cross-country skiing trails. To make artificial snow requires 0.6–2 kWh/m³ for pumps and compressors. The density of artificial snow is 0.4–0.5 kg/litre (dm³), indicating the need for water. On top of this, pumps and compressors produce significant noise pollution.

6.4 WATER NEEDS FOR ENERGY

Water is absolutely essential to produce, distribute, and generate energy. The use of water for energy is competing with demands in other industries, in agriculture and for domestic use. Water will almost certainly be more costly and valuable every year, as value added processes in energy and industries require increasing amounts of water. The 'water footprint' (the amount of water consumed to produce a unit of energy) of different methods of fuel production shows how water consumption for operations making primary energy carriers available vary from

> 85% of global water withdrawal is used for thermal power plant cooling

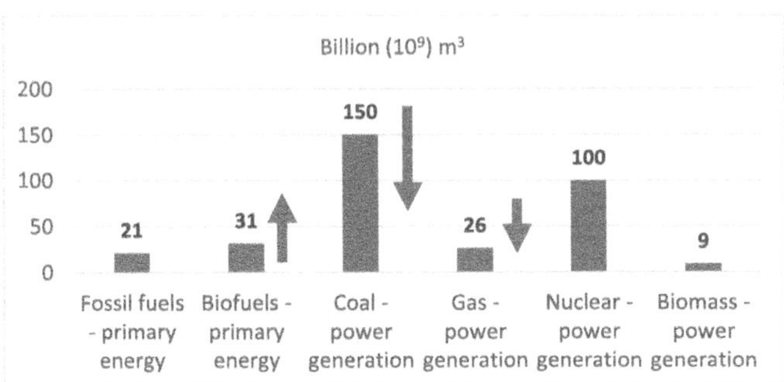

Figure 6.12 Global water withdrawal for different fuel types (IEA data 2016). The arrows indicate major predicted changes for the next decade. Water withdrawal is expected to increase significantly for biofuels and decrease for fossil fuel power generation.

fuel to fuel. Also, for electric power generation, there are different water needs for different fuel sources.

The global need for water for energy depends on the fuel type. Figure 6.12 shows the global *withdrawal* of water for energy extraction and electric power generation. The total global water withdrawal is 338 billion m³ and around 85% is used for thermal power plant cooling. The IEA has predicted how the water use will develop until 2030, indicated with arrows in the figure. Water withdrawal is assumed to increase for biofuel primary energy but will decrease for coal and gas power generation.

Figure 6.13 illustrates the corresponding *consumption*[205]. The total global water consumption is 47 billion m³ and energy extraction consumes around 70% of the water. The direction of the predicted water use is indicated in the figure. Like water withdrawal, water consumption will increase for biofuel primary energy but will decrease for fossil fuel extraction and

> 70% of global water consumption is for energy extraction.

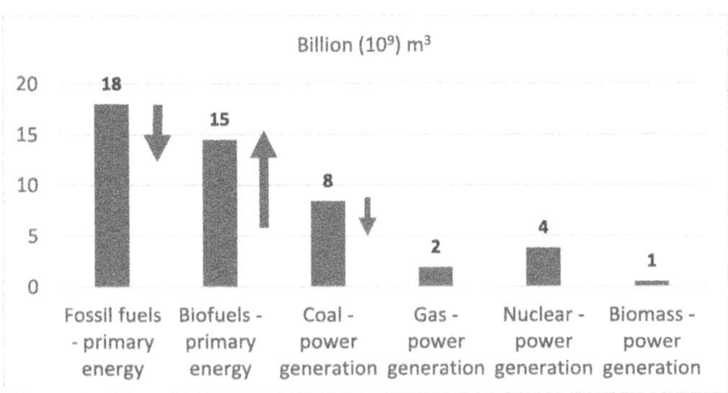

Figure 6.13 Global water consumption for different fuel types (IEA data 2016). The arrows indicate major predicted changes for the next decade.

coal power generation. Water consumption for other power generation types is not expected to change significantly.

It may be a surprise that biofuel is not as green as many of us think. The water to grow the crop and to extract the biofuel, such as ethanol or biodiesel, will require a huge amount of water.

6.5 FOSSIL FUEL PRODUCTION – IMPACT ON WATER AND ENVIRONMENT

The world's primary energy demand still depends hugely on fossil fuels as they supply more than 80% of the total (Figure 6.14). This number unfortunately has remained the same for more than a decade, despite the impressive development of renewable electrical energy. Transportation still depends mostly on petroleum. While coal production is levelling off (Figure 6.2), natural gas production is replacing the coal to a large extent. Admittedly, this will decrease the CO_2 emissions, but gas is still a fossil fuel.

Fossil fuel energy infrastructure is highly dependent on water availability. Oil and gas facilities employ water flooding to increase production. Around one-third of global refining capacity is currently located in high water stress areas, and this share is set to increase. Decreasing availability of water already affects refinery throughput in countries such as India, Iran, Iraq, and Venezuela.

6.5.1 Tar sand operations

The tar sand found in Alberta, Canada holds the world's third biggest oil reserves. Tar sand along the Athabasca River was detected as early as in the end of the 18th century, and then it was burned like coal. The tar sand exploration in recent decades that has taken place around Fort McMurray is one of the largest industrial developments in the world. The

> Tar sand oil is the largest export from Canada, causing 25% of total emissions.

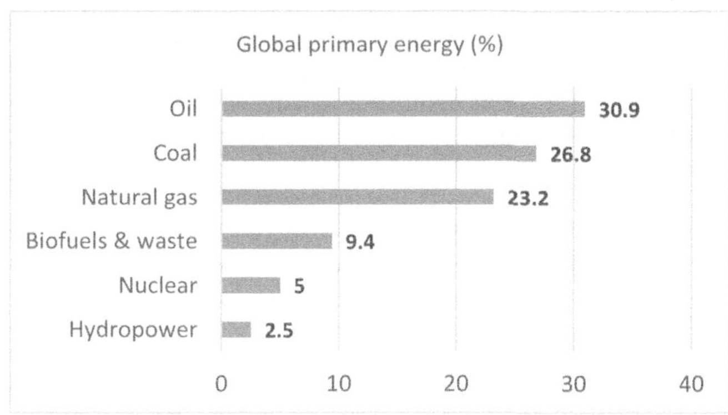

Figure 6.14 Primary energy supply sources worldwide in 2019. Fossil fuels contribute with almost 81%.The total primary energy supply is 607 exajoule (10^{18} J) or $169 \cdot 10^3$ TWh.

oil from tar sand is also one of the most polluting fossil fuel operations. The operation is gigantic with a production of around half a million m^3 of oil per day. This makes the tar sand oil the largest export from Canada and the nation's largest source of greenhouse gas emissions, around 25% of the total emissions. Canada is now the fourth largest oil producer. The negotiations in Glasgow 2021 have not influenced the oil production in Alberta, and there is no plan to lower the production.

The oil in Alberta is found in semi-solid form mixed with sand and water and is usually referred to as crude bitumen. This is a sticky, black, tar-like form of crude oil which is so thick and heavy that it must be heated or diluted before it will flow. The reason that the tar sand oil is more polluting is that the bitumen oil has a lower hydrogen-to-carbon ratio (about 1.5 on atomic basis) compared to conventional crude oil. Conventional oil has 2 H atoms for every C atom while methane has 4 H atoms for every C atom. The bitumen has also a higher content of sulfur, nitrogen, and metals, such as nickel and vanadium. A typical ore in the Athabasca oil sands contains 8–14% (weight) of bitumen and 3–5% (weight) water.

Oil sand exploration requires at least three times more water than conventional crude oil exploration. The oil sand consists of layers of sticky, tar-like bitumen mixed with sand, clay, and water. In many deposits some 30 metres of soil must be stripped off to reach the oil sand. The sand is delivered to an extraction plant where the bitumen is separated from sand in a hot-water wash, sometimes with caustic soda ($NaOH$) to facilitate the separation of bitumen from solids. The bitumen rises to the top of the wash and the sticky load is sent to an upgrading facility that converts it to synthetic crude oil. The biggest part of the recoverable oil is located deeper. When the bitumen is located too deep to be strip-mined, the industry melts it 'in situ' with copious amounts of steam, thus decreasing the bitumen viscosity, so that it can be pumped to the surface. Actually, 80% of the potentially recoverable bitumen is in deposits deeper than 60 metres. These deposits cover an area much larger than the mines, some 130 000 km^2.

Sand, water, and bitumen residues are finally piped to a 'pond' where the water is extracted, cleaned, and reused in the mines. The 'pond' is rather lakes of toxic mine tailings that serve as a settling basin. The current tailing ponds in operation occupy an area of around 250 km^2. The volume of tailing ponds will increase at an alarming rate with the current size of operations. To get 1 m^3 of synthetic crude oil requires around 11 tons of oil sand at 90% recovery, implying that *one day* of operations requires some 5 Mt of oil sand.

The Canadian National Energy Board (NEB) has calculated that to produce 1 m^3 of synthetic crude oil some 2–4.5 m^3 of water is needed. According to government figures from 2019, some 0.6 million m^3 (or 7 m^3/second) of water is drawn every day from the region's rivers, lakes, and aquifers. The wastewater contains a lot of hydrocarbons, naphthenic acids, and carcinogenic heavy metals. Nitrogen and sulfur

> Extracting 1 m^3 of tar sand oil pollutes 2–4.5 m^3 of water.

oxides are emitted into the air. The settling will take several years, probably several decades, so reclaiming the water is a tremendous challenge.

The dams create not only a water pollution risk for the groundwater but also for the Athabasca River close by. Millions of migratory birds arrive in northern Alberta each year and naturally they are attracted by the toxic water surface. The birds are meant to be scared off by air cannons that sound like gunshots blast several times every minute. Still there are tens of thousands landing in the toxic water each year[206].

It takes a lot of energy to extract the synthetic crude oil. The sand emerging from the mining is mixed with more water that must be heated to between 45°C and 80°C to wash out the bitumen. The heating decreases the viscosity of the bitumen. At the so-called upgraders the bitumen gets heated again to about 500°C and compressed to more than 100 bar (10 MPa) in order to crack the complex molecules into the light hydrocarbons that we use in our cars. This synthetic crude is then transported to conventional refineries for a final transformation into fuels. Most of the energy to heat the water or make steam comes from burning natural gas, which is the cleanest burning fossil fuel. In other words: the oil sand industry is wasting the cleanest fossil fuel to make the dirtiest fuel. According to Canada Energy Regulator, the oil sands gas consumption was 30% of natural gas use in Canada in 2018.

> 30% of all natural gas in Canada is used to produce tar sand oil

> The tar sand oil industry is consuming the cleanest fossil fuel to make the dirtiest fuel.

The tar sand operations have caused a lot of damage to Indigenous groups of people. The concept of 'ecocide' – systemic, long-lasting environmental devastation – has been used to recognize the damage to the ecosystem, also including the Indigenous people's cultural, spiritual, and subsistence ties to nature. Numerous 'priority pollutants' such as mercury, lead, nickel, and other heavy metals, are found in the river downstream of the oil sands development. Elevated levels of polycyclic aromatic hydrocarbons (PAH) had already started rising in the 1960s when the oil sand operations began. These compounds are causing cancer. Higher concentrations of toxic pollutants are also found in the flesh of moose, ducks, and muskrats. Correlation has been found between cancer rates and working in the mines and consuming larger quantities of traditional foods, particularly fish. The air pollution is another high health risk. Some 150 nations have signed a Treaty against Tar Sands Expansion[207].

Transportation of the tar sand oil via pipelines to the USA is creating a lot of controversy. The risks for pipeline leakages cannot be eliminated and the existing Enbridge pipeline has had several leaks during recent decades. Oil spills from tar sand are more devastating than leaks from conventional oil. The bitumen is diluted (known as dilbit) in order to be pumped and it is particularly difficult to clean in aquatic environments. The mixture evaporates rapidly and separates from the bitumen. The heavier bitumen then sinks instead of staying at the surface, like most conventional oils.

Sometime in the future, when fossil fuels are no longer being used, a clean-up process has to start in Alberta. It has been estimated that the cost would exceed US$100 billion. So far, the Alberta government has secured Can$730 million from companies for the clean-up.

6.5.2 Hydraulic fracturing

Hydraulic fracturing or fracking has seen an enormous development, particularly in the USA It has been described in more detail earlier[208], so here we present a short overview of the latest development of the water and environmental impact of fracking.

The two primary water issues associated with fracking are:

> Fracking fluid is contaminated. Problematic to treat in conventional wastewater treatment systems

- The use of a large amount of freshwater to accomplish the fracking
- The necessity of protecting underground water tables and surface water resources from contamination by fracking fluids and/or migrating gas deposits.

The fracking fluid becomes contaminated and is most difficult to treat in conventional wastewater treatment systems. There is a risk of leakage from improperly treated produced water and fracking fluids from flowback into the soil and water table.

Often shale gas is found in dry regions, so fracking takes place in many areas where water scarcity is a reality. While the economic driving forces are huge, the environmental consequences of shale oil and shale gas and the impact of hydraulic fracturing on air and water quality are intensely debated. According to 2016 data[209] for the USA almost 60% of the more than 100 000 wells that were hydraulically fractured are located in regions with high or extremely high water stress, primarily in Texas, Colorado, Oklahoma, and California. Texas alone has around 50% of all the US wells. The average water use per well doubled between 2011 and 2015 and reached 20 000 m^3 per well.

The fracking fluid contains chemicals of concern as well as radionuclides. Biocides and certain petroleum products that are present in fracturing fluid are particularly hazardous chemicals that may cause health risks that range from rashes to cancer. Over time the fracking wells produce much more wastewater than conventional oil and gas extraction, often 10 times more. Therefore, water

> Several chemicals in the fracking fluid are highly toxic

management costs will rise. The amount of wastewater that is returned to the surface is almost nothing in some places and almost all of it in other regions, depending on the geology. Most often the 'used' water is not treated but injected in deep wells. This has been done in almost 700 000 wells only in the USA. There are incidents of leakage into aquifers as well as surface water contamination. Deep well injection is prohibited within the European Union.

Methane contamination has been a common complaint among people who live near natural gas drilling areas, such as the Marcellus Shale in the Appalachian Basin, Pennsylvania. There is a natural background leakage into the drinking water in the area. However, researchers found a chemical signature to make the connection to fracking. Water samples containing high concentrations of methane in areas with fracking operations had higher concentrations of dissolved iron and sulfate (SO_4^{2-}), while samples from non-fracking areas had normal iron and sulfate concentrations[210].

China has the world's second largest technically recoverable shale-gas resources. Unlike in the USA, the most attractive Chinese reserves are located in remote, mountainous areas. Shale resources lie deeper than in the USA, around 3500 metres. The primary focus is on the south-western Sichuan province, holding around half the national shale reserves. China is aiming to produce more natural gas from the shale resources to cut its reliance on coal. Also, in China most shale-gas reserves are located where there are water shortage issues and there is a great risk of increasing water-energy tension.

6.5.3 Flaring

Natural gas is deliberately burnt by the oil companies. The gas bubbles up alongside the far more valuable oil. Technically the gas can be captured and utilized. With less economic incentive to capture it, the drillers treat the gas as waste and simply burn it. Seven countries – Russia, Iraq, Iran, the USA, Algeria, Venezuela, and Nigeria – produce 40% of the global oil each year but they contribute with 65% of the global gas flaring[211,212].

This corresponds roughly to the natural gas demand of Central and South America. There are great discrepancies in flare intensity. Venezuela flared 40 m³ of gas per barrel (or 250 m³ gas per 1 m³ of oil) while the USA flared 3 m³ per barrel (or 19 m³ gas per 1 m³ of oil). The flaring praxis results in severe air pollution, including CO_2, methane, and soot. The methane emissions from flaring are particularly harmful.

> Flaring praxis causes air pollution, CO_2, methane, and soot

Nigeria has been particularly vulnerable to climate change, but the country's oil industry is making things worse. Flaring is illegal in Nigeria and the Nigerian government has promised to end flaring by 2030. Yet, about 2 million people live within 4 km of a gas flare, creating toxic pollution. The inhabitants of the land suffer from abject poverty. The temperature around a gas flare can reach intense levels. Some people use this heat in a risky operation to dry food.

Gas flaring has been falling globally and great efforts have been made in the USA, in oil fields in East Siberia, and in Nigeria. In total, gas flaring is still releasing over 400 Mt of CO_2 equivalents or of the order 1% of total global emissions. However, methane leakage does not seem to be prioritized in Russia.

Gazprom announced in 2020 that they release methane corresponding to 25.5 Mt of CO_2 equivalents.

6.5.4 Coal

Coal is still a large provider to the world's primary energy, around 31% and contributes around 40% of all CO_2 emissions (Figures 6.2 and 6.3). Coal burning is a major cause of urban smog. It is the dirtiest (and cheapest) of the fossil fuels. Furthermore, it is used in the production of 70% of the world's steel.

> Except for power generation, coal is still essential in steel production

Coal mining has an enormous impact on both land use and on surrounding water quality. Mining and refining coal requires water at various stages. Refining coal means that the coal is washed, and additives are mixed with the feedstock coal in order to reduce emissions when burned. Estimates show that approximately 0.16 litres of water is needed per MJ, or 4 litres of water per kg of coal.

The top seven producers are shown in Figure 6.15a. They provide 91% of the world's production. Figure 6.15b is the ratio between domestic consumption and production. India and China are using more coal than they can produce, while Australia and Russia export most of their coal. China has a dominating influence on the global coal market and China's electric power generation accounts for one-third of the global coal consumption, while the overall coal use is around half the global use.

The production numbers give an indication where the largest environmental impact of coal mining takes place. The consumption, on the other hand, gives a hint where the largest environmental consequences of coal burning are located.

According to IEA the final investment decisions (FIDs) for coal-fired power capacity increased slightly in 2020 to reach 20 GW, primarily concentrated in

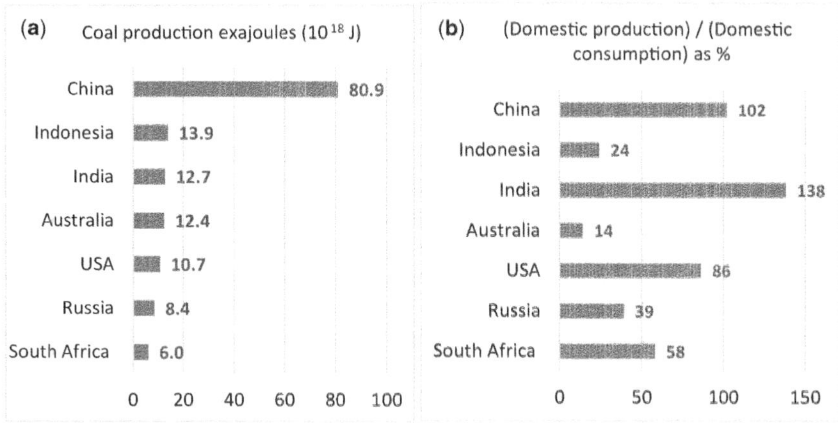

Figure 6.15 The top seven coal producers in 2020 (in exajoules, 10^{18} J) (a). The ratio (%) between domestic coal consumption and coal production[213] (b).

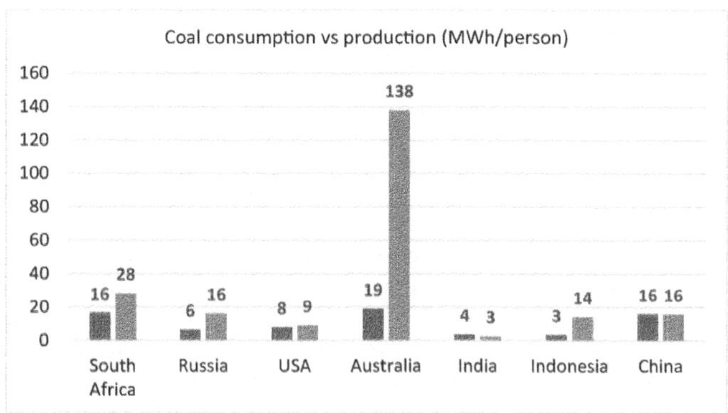

Figure 6.16 The seven largest coal producers and their domestic consumption per person (left bar) and production per person (right bar).

Asia. However, the longer trend is shrinking investments. FIDs are now 80% lower than what they were 5 years ago. However, the future of coal will largely be determined in major economies in Asia where electricity demand is still growing in many cases.

Figure 6.16 illustrates the coal consumption and production respectively among the seven top producers. Australia's production per person is significant compared to the rest of the top producers.

The coal consumption to produce 1000 MW of electric power is huge. Some 8 tons of coal has to be burned every minute, requiring about 100 train cars or around 12 000 tons of coal every day.

Virtually every stage of the coal lifecycle – from mining to processing to burning – can impact water quantity and quality. Coal mining has a devastating impact on water quality. An important consequence of coal mining is the so-called acid mine drainage (AMD), caused by the contact between coal surfaces, air, and water. The pyrite (iron sulfide, FeS_2), also known as 'fool's gold', oxidizes to ferrous sulfate and forms sulfuric acid (H_2SO_4). This acid will be mixed with drainage water from the coal mine. Many discharges have a pH less than 3.0 and the water generally has high concentrations of arsenic, copper, lead, manganese, aluminum, and sulfate. The costs to clean up the AMD pollution from abandoned mine-land sites are substantial.

> Virtually every stage of the coal lifecycle – from mining to burning – can impact water quantity and quality.

Methane can be absorbed by the coal deep in the ground and will be released during the mining process as a result of decreased pressure and the fracturing of the strata during the mining process. As described in 4.9, satellite observations can reveal any methane leakages. It is well known that pockets of hazardous gases are a serious threat to coal miners.

Surface mining is a most destructive way of coal mining. Not only does it change the appearance of the landscape, but it also affects streambeds. Wherever it occurs in the world, it eliminates existing vegetation, destroys the genetic soil profile, displaces or destroys wildlife and habitat, degrades air quality, alters current land uses, and to some extent permanently changes the general topography of the area mined. The impoundment ponds – sometimes called slurry ponds, sludge lagoons, or waste basins – often remain.

After coal is mined it is typically washed with water and chemicals to remove impurities. South African calculations show that coal washing, extraction, dust control and evaporation may require as much as 430 litres of water per ton of coal produced. US numbers are of the order 75–150 litres per ton. The coal slurry must be stored in ponds that can leak or fail. These water resources are rarely cleaned up after the mining operations have closed and will create a permanent degradation to agricultural productivity. Furthermore, the cost for the water is usually much lower than the cost for households.

Coal combustion in power plants generate waste products, primarily coal ash, containing radioactive elements, minerals, and heavy metals like arsenic, mercury, cadmium, and lead. Some 80% of the coal ash is fly ash, which is respirable and has a primarily a glass chemical structure. Coal ash is mostly stored in impoundments containing a mix of fine powder and sludge, most often located near low-income residential communities.

Coal mining and burning has a devastating consequence for air quality and public health, in particular emissions of local pollutants, particularly sulfur dioxide (SO_2), nitrogen oxides (NO_X), and particulates.

UN Secretary General António Guterres has clearly stated that 'accelerating the global phase-out of coal is the single most important step' toward meeting the Paris Agreement's goal.

6.5.5 Oil accidents and oil pollution

Oil accidents have been widely published elsewhere and the impact of oil extraction and oil transportation is well-known. My previous book[214], as well an earlier publication[215], describe major disasters in the Mexican Gulf in 1979 and 2010 and in Alaska in 1989, and the human and ecological cost of oil exploration in the Niger Delta in Nigeria. Here we highlight another impact, the human cost of oil by a short account of more recent Niger Delta experiences.

Many international and national legislations and conventions, including the UN SDG6, have underlined safe water as a human right and a prerequisite for the realization of other rights, such as the right to health and food. The Niger Delta in Nigeria holds massive oil deposits, which have been extracted for decades. The suffering of the population because of the oil exploration has been observed internationally. UNEP[216] and organizations like Amnesty International have alerted the world about how oil exploration has created huge suffering for the locals. As expressed by UNEP: 'most members of

> Oil leakages are not always accidents, but often caused by negligent management.

the current Ogoniland (eastern part, Niger Delta) community have lived with chronic oil pollution throughout their lives'. Estimates of the catastrophe vary but put the number at over 10 000 spills since the oil industry began operations in the late 1950s. UNEP has put the *annual* average volume spilled into the Delta environment 18 000–44 000 m³, or total oil spills 1.4–2.1 million m³, making the Delta one of the world's most oil spill affected areas. The accumulated oil spills correspond to a 1.5–2 mm oil cover all over the 1000 km² Ogoniland in the Niger Delta.

A major water-related oil spill in 2008–2009 case was litigated in UK in 2011 and lasted until 2014. Amnesty International[217] has documented the Bodo oil spill justice struggle. The UK£55 million (US$83 million) legal settlement between Shell and Bodo is one of the most significant legal settlements ever reached against a transnational corporation. By comparison, BP paid around US$65 billion in clean-up costs and damage payments for the Mexican Gulf spill. Fundamentally, the Bodo legal settlement was a victory driven by a series of limnology datasets and advocacy. Research data from 2006 onwards provided pre-impact baseline and post-impact hard evidence for the UK lawsuit. The case was brought to the attention of the lawyers in London who ultimately pursued it in the British legal system for 4 years, requiring lengthy and complex negotiations. By coupling limnology with human rights, it was possible to secure a historical environmental justice victory.

The ongoing Bodo clean-up was meant to be the world's largest mangrove remediation project. It involves free-phase oil removal, remediation, and restoration (i.e., planting of about 1000 ha of degraded mangroves). The clean-up project was initiated in 2013 through the Bodo Mediation Initiative (BMI), a multi-stakeholder clean-up management platform created by the Dutch Embassy and NACGOND, a coalition of civil society groups that seek to address the environmental degradation associated with oil spill, gas flaring, and illegal oil storage in the Niger Delta.

In the Niger Delta, decades of poorly regulated oil production and management of the ancillary impact on the environment continue to degrade the quality of water resources. Persistent oil spills and poor management of oil contamination of water systems impact fundamental human rights of access to safe water resources, health, and food, which are inextricably interlinked. Pollution of water resources affects a gamut of ecosystem goods and services. For example, oil spills directly compromise the water quality of an estuary and kill mangroves that help to purify and maintain the water quality by trapping and settling suspended particles in the water, which further build up the local sediment (accretion) to support mangrove plants, which in turn, supports fish breeding and nurseries.

There is urgent need to forestall further pollution of water resources, and remediate and restore impacted water systems, especially in light of the UN Decade for Ecosystem Restoration. A crucial experience from the long struggle between the local population in Ogoniland and the international oil companies is that high local competence and financial support is needed to continuously monitor and report water quality.

6.6 BIOFUEL PRODUCTION

The poor part of the world population depends on traditional biomass such as firewood, charcoal, crop residues, and dried animal waste. Biomass is mostly available and affordable, especially for cooking and space heating. However, biomass used in small-scale appliances is rather inefficient and highly polluting.

The so-called first-generation biofuels include ethanol and biodiesel. Ethanol is an alcohol and can be produced from corn, sugar cane, sorghum, potatoes, and wheat, as well as from cornstalks and vegetable waste. Ethanol used as fuel for cars is mostly sold as a gasoline additive or as E85 (85% ethanol and 15% gasoline). The chemical name for biodiesel is methyl esters and is intended for diesel engines. It is made from natural oils such as animal fats or vegetable oils.

> First generation biofuels are competing with food.

The typical feedstocks used for the manufacturing of the fuels are food crops rich in starch, such as sugar cane, sugar beet, and starch-bearing grains (e.g., corn (maize) and wheat), oil crops (e.g., canola and palm) and even animal fats. Ethanol from corn is unsustainable. In the USA 40% of the corn harvest – one of the most important agricultural products in the country – goes towards ethanol.

Ethanol became the premier renewable fuel, in particular after the Renewable Fuel Standard (RFS) was passed by the US Congress in 2005. This initiated the world's biggest biofuels programme, and the US biofuel production was 47% of the global output over the last decade. The ethanol programme has been seriously questioned in a new study[218]. The study found that the RFS programme increased corn prices by 30% and the prices of other crops by 20%. Corn ethanol has not decreased the demand for gasoline. Even worse, grasslands and forests have been converted into croplands, thus releasing carbon in the process. The result is also increased fertilizer use, water pollution, and habitat loss. The authors find that the ethanol programme may have even increased the greenhouse gas emissions compared to gasoline.

Second generation technology is developing with raw material from non-food cellulosic biomass and hemicellulose, the fibrous material that makes up the bulk of most plant matter. It is generally produced through two methods:

- Biochemical: hydrolysis and fermentation of woody or fibrous biomass, and
- Chemical and thermochemical processes.

In the biochemical conversion, the lignin is separated in biochemically and can be used separately for heat and power production. In the chemical processes the lignin is also converted into syngas. The biochemical processing produces only ethanol. The thermochemical processes, on the other hand, can produce a range of fuels with different properties. Biodiesel can be taken from oil-rich plant seed, such as soybean or jatropha.

The third generation of biofuels are generated from cyanobacterial, microalgae, and other microbes. This is considered a promising approach to meet global energy demands. The use of photosynthetic organisms as a source of biofuel is based on the fact that atmospheric CO_2 serves as a source of carbon and sunlight serves as the energy source. In this way, the basic molecule substrate is created for the production of bioethanol and biomethanol[219].

Biogas for energy is a proven and widely used global source of energy, not only in low-income countries but also in many highly industrialized countries. Using biogas means that firewood use can decrease. Furthermore, savings include chemical fertilizers and a reduction in illnesses associated with indoor air pollution. From a gender perspective, biogas has contributed to less cooking time for women, and less time and pain for women and children collecting heavy loads of firewood[220].

6.6.1 Water use

Water is essential to produce bioenergy and to produce the necessary amounts of agro fuel, agriculture requires the input of freshwater for crops. The more a decarbonization pathway depends on biofuel, the more water it consumes. Compared to fossil fuels, biofuels are a larger source of demand for both water withdrawals and consumption and its relative water use is expected to rise, as demonstrated in Figures 6.12 and 6.13.

The amount of irrigation water used to grow agro fuel varies significantly from one region to another and depends naturally on climatic conditions, the farming methods, and the processing technology used. This is similar to any food production.

The water requirement for biofuel production is discussed in detail elsewhere[221], so here we summarize some key interactions between biofuel, water, food, and energy. Transportation is a major user of biofuels.

Water scarcity, rather than land scarcity, may prove to be the key limiting factor for biofuel feedstock production in many contexts. The largest component of water use associated with bioenergy is the cultivation of feedstocks. Pollution of water by agro-chemicals can also be characterized indirectly as a 'water use' since it may reduce freshwater availability by contaminating water resources. Runoff containing fertilizers, pesticides, and sediments (surface and groundwater) will contaminate the water and the refining process will produce wastewater.

From a global perspective the overall use of water for biofuels production is modest compared to water used for food production. It is estimated that less than 2% of agricultural irrigation is used for biofuel production. Naturally, in countries where water is already scarce, growing biofuel crops will aggravate existing problems. For example, the situation looks serious in both China and in India, where programmes to boost biofuel production have been initiated. Both countries have already serious water scarcity problems.

> Water scarcity is a limiting factor for biofuel production.

The water requirement for crop growth is reflected in crop water transpiration. The biomass production is close to linearly related to the transpiration. A

Table 6.1 Energy balance for different biofuels of first-generation biofuels, expressed as the ratio between biofuel energy unit and the required input energy[224].

Biofuel	Energy ratio
Corn ethanol	1.3
Sugarcane ethanol	8
Biodiesel	1.9– 2.5

typical estimate is that 200 litres of water transpired is needed for every 1 kg of biomass grown[222]. There are further water losses from evaporation, water runoff, and leaching.

The conversion and processing of feedstocks require much less water than the crop growth. Still, biofuel refineries need significant amounts of water, mostly as steam for fermentation, compared to oil and gas refining. For example, to produce 1 litre of ethanol would require the following volumes of water:

- Fermentation of corn grain: 2–4 litres
- Biological cellulosic conversion: >9 litres
- Thermochemical gasification: <2 litres of water.

To produce biofuel requires energy. Ethanol production with existing technologies is most energy efficient from sugar-based feedstocks, such as sugarcane and sweet sorghum. Producing ethanol from corn grain requires high energy both to grow corn and to convert it into ethanol. The net energy balance is low, Table 6.1. Soybean diesel has a better energy balance. The reason is that soybeans create long-chain triglycerides that are easily expressed from the seed[223].

There are negative impacts on the environment producing corn and soybeans for biofuels. Leaking fertilizers and pesticides can create harm to the environment. It has been noted that corn production creates a larger environmental burden than soybean production.

6.6.2 Burning wood

Burning wood is based on the false assumption that energy from forest biomass is carbon neutral when it is actually a major emissions source. The European Academy of Sciences Advisory Council (EASAC) has expressed a strong opinion about this and has focused on the net climate impacts of conversions for electricity generation from coal to forest biomass, which attract billions of euros in subsidies[225].

> Burning wood is not renewable in the short time perspective.

When a power station switches from coal to wood pellets, a significant amount of CO_2 compared to coal is released at first. The reason is that biomass has a lower energy content than coal. Moreover, the supply chain is complex. Later, vegetation regrows, and the extra CO_2 will be reabsorbed from the atmosphere. However, this time is far beyond the time we have available to meet Paris Agreement targets of limiting warming to 1.5–2°C. The long-term emission of wood is favourable, but, as EASAC reminds us, 'the devil is in the details'. The time that it takes to regrow the wood, the *carbon payback period*, depends on a lot of factors, but can be several decades. We need to reduce atmospheric CO_2 levels much quicker.

6.7 ELECTRIC POWER PRODUCTION

Figure 6.5 shows the global electrical energy mix in 2020 and clarifies the dependence on fossil fuels which provide 63% of the electric power generation.

To compare the lifetime costs of generating electricity across various generation technologies, the economic measure of Levelized Cost of Electricity (LCOE) is used. The calculation includes the following parameters:

- Total capital cost
- Fixed operation and maintenance (O&M) cost
- Capacity factor
- Useful life, and
- Discount rate.

A common LCOE calculation is presented regularly by Lazard's annual LCOE Analysis. The latest one is partly shown in Figure 6.17.

Another estimation is made by the International Renewable Energy Agency (IRENA). The key message is that the price decrease of wind and solar photovoltaic since 2010 has been dramatic, as shown in Table 6.2.

There is a difference in the economic structure of renewables and conventional electricity generation. The major cost for renewables is the investment, while the 'fuel' is free in terms of wind and sun. For conventional power, the fuel price is not only apparent, but it is also depending on the market and on geopolitics.

It is outside the scope of this book to evaluate different future energy solutions. Our aim is to indicate various interactions to and from various energy sources.

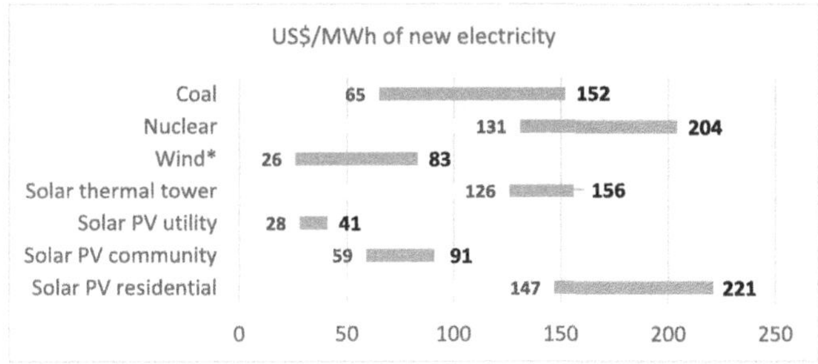

Figure 6.17 LCOE comparison between different electrical energy generation systems by Lazard (28 Oct, 2021). The costs are unsubsidized and not specific to any region[226]. Hydropower is not included.
*Both onshore and offshore.

Table 6.2 Levelized costs in 2020 for different electric power sources, expressed in US$ for every new MWh.

	US$/MWh	Change (%) 2010–2020
Hydropower	44	+15
Solar photovoltaic	57	−85
Concentrated solar power	108	−68
Offshore wind	84	−48
Onshore wind	39	−56

The fossil fuel cost range is in the order 60–150 US$ for every new MWh[227].

6.8 HYDROPOWER

As noted in Figure 6.6, hydropower has grown significantly in the world during recent decades. From a power grid operation point of view, hydropower is an excellent source of energy. The electrical power output from a hydropower plant can be changed within fractions of a minute making hydropower the preferred source for frequency control as well as a flexible compensation for variable production from wind and solar power.

The obvious relationship and source of conflict between water resources, ecosystems, land use, and hydropower will create challenges. It is discussed in more detail elsewhere[228]. In section 5.4 we also described the impact of rising temperatures and increasing use of water on hydropower generation. The impact associated with dams in a cold climate is significantly different from that of dams built in arid or tropical areas.

6.8.1 Impact of reservoirs

Multipurpose dams have been built for irrigation, for storage of water to compensate for seasonal variations, and to avoid severe flooding. Hydropower then comes as an extra advantage of the water reservoir.

6.8.1.1 Evaporation

When water is stored in a basin instead of flowing in a river in a warm country, the temperature will rise causing increasing evaporation. Climate change will amplify evaporation. Factors determining the amount evaporated – climate, reservoir design, and allocations to other uses – are highly site-specific and variable. This suggests that hydropower plants with large dam capacity at some sites can have some of the highest water consumption levels of any capacity type per unit of electricity generated. Run-of-river hydropower plants, however, store little water, leading to evaporation losses that are near zero.

Elementary calculus can qualitatively explain that evaporation water loss is larger for a large surface/volume ratio of the basin. Evaporation in warm countries can be significant. For example, Lake Nasser upstream of the Aswan Dam in Egypt loses about 3 metres every year due to evaporation, causing a devastating reduction of the flow rate of the River Nile. This will also influence the hydropower generation and irrigation capacities. A typical reservoir in India loses about 1.5 metres per year and in dry areas of Australia the loss can be as much as 3 metres per year. An extreme example is the Akosombo dam in Ghana, a giant reservoir having an area of 8500 km². It loses more than 2 metres per year which corresponds to about half the capacity of the Volta River in an average year. It is argued that evaporated water is not consumed but re-enters the hydrological cycle as precipitation. However, the precipitation may not compensate for the evaporation where the water was used previously. The control of that amount of water has been lost due to the evaporation.

> Reservoir evaporation is a serious problem in warm countries.

6.8.1.2 Public health

The increasing temperature caused by damming and flood control can have public health consequences, for example increases in diseases such as schistosomiasis and malaria. Bilharzia or schistosomiasis is a water-borne disease, currently infecting more than 200 million people. It is a common problem in tropical areas that after the building of dams and reservoirs both malaria and bilharzia have increased. Schistosomiasis is transmitted best in water which is not flowing like in lakes and rivers after damming.

As noted in the prologue Georg Borgström had already warned about this problem in the 1960s. Another issue, often overlooked, is that the local people over the centuries have developed a flood-dependent agriculture with a varied food production. When the positive effects of the flooding are gone, then the food production has in fact decreased with resulting malnutrition. The reason is often that the food production consists of monocultures grown on irrigated land.

6.8.1.3 Fishing and biodiversity

An apparent topic for public discussion concerning hydropower is that the natural beauty of nature that is flooded by the dam, the lost grandeur of the free-flowing river water is worth a great price, but how much? The land that used to be a great touristic attraction is now lost. Sometimes a pretty lake is created but in most cases the value of the hidden nature was higher. The economic value of tourism should not be underestimated.

> Increasing temperature caused by damming and flood control can have public health consequences

While the expansion of hydropower is expected to double, it could reduce the number of our last remaining large free-flowing rivers by about 20% and pose a serious threat to freshwater biodiversity. The hydropower boom occurs primarily in emerging economies in South America, Southeast Asia, and Africa. These regions also hold some of the world's most important sites for freshwater biodiversity. A new database has been developed and announced in October 2014 to support decision making on sustainable modes of electricity production (www.freshwaterbiodiversity.eu).

Obviously, a dam will alter the flow of a river and will disrupt the migration and spawning of fish. It is common that lakes downstream of a large dam will shrink since they no longer absorb the river overflow. China has declared in the 13th Five Year Plan to 'strictly control the expansion of small hydropower stations' to protect the environment. This would include the plan to improve or remove some 40 000 small hydropower stations. Naturally, this is easier said than done. To remove coal powered plants and increase fossil-free energy while removing or upgrading hydropower stations is not trivial.

Damming and flood control may cause declining fish catches and loss of freshwater biodiversity. Two cases, among several, are mentioned for illustration. In Egypt, the massive Aswan Dam has caused the fertile Nile Delta to shrink. There are significant consequences for both fishing and agriculture[229]. There is a massive hydropower development in the Mekong River Basin in Southeast Asia, which is the site of the biggest inland fishery in the world. Fish migration routes between the river's downstream floodplains and upstream tributaries can be blocked by the hydro dams. There are alarming reports that the completion of several dams on tributaries will have catastrophic impacts on fish productivity and biodiversity[230].

6.8.1.4 *The risks of dam failures*

Many dams being built over the years to provide hydropower have not been built with an environmental and safety assessment. China has most hydropower dams in the world with an installed electric power capacity of 385 GW. The Yangtze and its tributaries alone had more than 24 000 hydropower stations by the end of 2017 and at least 930 of them were considered constructed with an environmental assessment. Many old dams are not safe, in particular during summer floods. More than 3500 reservoirs burst during the 60-year period 1951–2011, according to China's Ministry of Water Resources. The Banqiao dam and several smaller dams in Henan province broke in 1975 because of heavy rains, causing the death of 240 000 people.

6.9 THERMAL ELECTRIC POWER GENERATION SYSTEMS

Electric power plants can be characterized by both carbon and water footprints. The *carbon* footprint per generated kWh is the highest for coal-powered plants. Natural gas will emit less, while the lowest is for nuclear plants. The *water* footprint for water-cooled thermal plants is typically high. Water limitations are getting increasingly difficult in many places. One out of three freshwater-cooled thermal power plants are located in areas of high water stress. The IEA predicts that the frequency of extreme heat events would double by 2050 compared with today – and they would be around 120% more intense. This will influence both the grid performance and the thermal plants and increase the demand for cooling[231]. IPCC predicts in its RCP (Representative Concentration Pathway) 4.5 scenario (an intermediate emissions scenario) that over 40% of freshwater cooled thermal (coal, gas, nuclear, and oil) power plant capacity is projected to be in high-risk areas by 2040. This highlights the urgent need to enhance the resilience of energy systems to climate change.

6.9.1 Nuclear power

With an increasing pressure to shut down fossil fuel electric power plants, there is a push to expand nuclear power as a base load solution for electric power generation with an increasing fraction of variable power, such as wind and solar.

The nuclear industry is well established, and today there are almost 450 units in operation in the world, producing some 10% of the global electricity. However, at least 150 of the nuclear reactors are to be decommissioned during the next decade. Most of the reactors were designed for a 40-year useful life. Almost half of them are 31–40 years old. Since a nuclear reactor is large, the shutdown of single units will be noticed on a national scale. This was noticed after the Fukushima disaster in 2011. Recently six plants have been shut down in the USA, Sweden, Russia, and France.

The completion of new reactors has not been easy. The Finnish Olkiluoto nuclear station has overrun by 9 years and the French Flamanville, using a similar technology, has overrun by 7 years. Lead times for nuclear designs and projects are long. Also, new regulations and political changes during the lifetime of a construction can add time and cost. Standardized design and construction may be hard due to the long construction time, and new designs in key components create unexpected problems and delays.

> *New* nuclear reactors cannot contribute to limit global warming within the next 10–15 years.

At the end of 2020, 50 nuclear reactors were under construction. In the years 2016–2020 there were 29 reactors constructed in the world, having the average construction time of 93 months or 7.75 years[232]. There were 44 reactors younger than 5 years on 31 December 2020. Given the long-time lag from decision to completion it means that the young reactors are a result of decisions taken more than a decade ago. This should be kept in mind in climate discussions. To realize the pledges made in COP26, energy generation must make significant changes before 2030. This is not possible to achieve with new nuclear power. However, nuclear and renewables are not in competition. Every zero-carbon power generation technology may be needed.

Expanding nuclear power is politically sensitive. The advantages are that the power plants do not emit CO_2 and can provide a base load, even if there is no wind and sunshine. Typically, the electric power demand is high on a cold winter day, and this often coincides with hardly any wind. The negative aspects include environmental impact of uranium extraction, the risk of serious accidents, and the difficult handling of radioactive waste. However, it should be noted that nuclear reactors are not designed to compensate for the variations of wind and solar production. This regulation is done by hydropower, when available.

One future scenario is a massive deployment of small, modular nuclear reactors designed to complement wind and solar technologies. That scenario includes nuclear power applied not just to electricity, but also to manufacturing hydrogen. The advantage would be that standardization could make the price lower. Several early-stage companies are testing these smaller, more modular nuclear generation technologies. However, no commercial plant has yet been built. To scale up a technology that is still not proven and make it commercially viable within one or two decades still looks like a dream. Nuclear solutions are too slow to be relevant to solve the climate crisis this decade. On the other hand, the 'not in my backyard' argument is relevant in all energy production.

The so-called fourth generation nuclear reactors are focus of the discussion. This is not a single type of design, but all of them have two things in common: a high operating temperature and a more demanding reactor core material challenge. In reactors today, fast neutrons from the fission are slowed down by the moderator (water or graphite) to split the U_{235} nucleus. Natural uranium contains only 0.7% U_{235}. Most of the uranium is U_{238} and this is not used. To split this nucleus, the neutrons must have a high speed. In other words, they are now allowed to be slowed down. As soon as water is present – as a moderator and as a cooler – the neutrons are slowed down. It can simply be understood as a billiard game. The neutrons have a mass like hydrogen atoms, and when two similar 'balls' collide they will lose speed. Therefore, the solution to keep the high speed of the neutrons is to use much heavier atoms to cool the system.

It is here that the tough material problems appear. Cooling candidates for fast reactors are liquid sodium (Na) (melting temperature: 97.7°C and boiling temperature: 882.8°C) and liquid lead (Pb) (melting temperature: 327.5°C and boiling temperature: 1750°C). The reactor core material will have to withstand high operating temperatures, intense fast neutron fluxes, and contact with corrosive coolants like sodium and lead. Specific alloys are being developed to help in combating these issues, but they still seem far off from full-scale implementation. Of course, it looks attractive if most of the uranium can be used as fuel and this is one of the key arguments for fast reactors. This is not a new argument: as a nuclear engineer in the 1960s I was involved in many discussions about fast reactors.

The IEA claims that the nuclear industry will need to nearly double in size over the next two decades for the world to meet net-zero emissions targets. The USA has accepted nuclear power as a solution for low-income countries like Kenya, Brazil, and Indonesia. Russia is pushing other countries to recognize its nuclear power plants as environmentally friendly. France is much in favour of nuclear energy (with 71% of electricity generated by nuclear) while Germany is much against it (with 11% of electricity from nuclear at the end of 2020 and shutting down all their nuclear reactors in the end of 2022).

James Hansen, the climate scientist, has been outspoken about nuclear power: *'There is no credible path to climate stabilization that does not include a substantial role for nuclear power... A major expansion of nuclear power is essential to avoid dangerous anthropogenic interference with the climate system this century... We've done the math and we can't power the world without nuclear energy.'*[233]

On the last day of 2021 the European Commission released a draft communication on a taxonomy of green economic activities: a guideline for sustainable investments[234]. According to the document certain nuclear activities will be called sustainable investments. Nuclear projects with a construction permit issued by 2045 would be eligible for private investments if they can provide plans for the management of radioactive waste and for decommissioning. The Commission will present the final text in January 2022. Still, the industry has not proven that the technology is safe, cost effective, and green.

6.9.2 Fossil-fuelled power systems

The CO_2 emissions from power generation are influenced by two factors, the carbon content of the fuel and the overall conversion efficiency. Huge efforts have been made to increase the thermal efficiency, but still, this is not sufficient to mitigate the climate change. The efforts include clean combustion technologies and combined cycle operations. Power plants that burn coal, natural gas, and petroleum fuels account for almost 100% of electricity-related emissions.

The amount of CO_2 produced when a fuel is burned is a function of the carbon content of the fuel. The amount of *energy* produced when the fuel is burned is main a function of the carbon and hydrogen content of the fuel. Natural gas consists mainly of methane (CH_4) and has a higher energy content relative to gasoline and coal and has a lower CO_2-to-energy content. Typically, natural gas emits around half the amount of CO_2 per energy unit, compared to various coal qualities.

6.9.3 Water requirements

Thermal power generation uses immense volumes of water. The installed electricity capacity worldwide in 2019 is shown in Figure 6.18.

The power capacity of fossil fuels and nuclear is about 65% of the total capacity. Given the variability of wind and solar, thermal plants provide a higher share of global electric power production. Per unit of energy produced thermal plants are the energy sector's most intensive users of water. Of the total

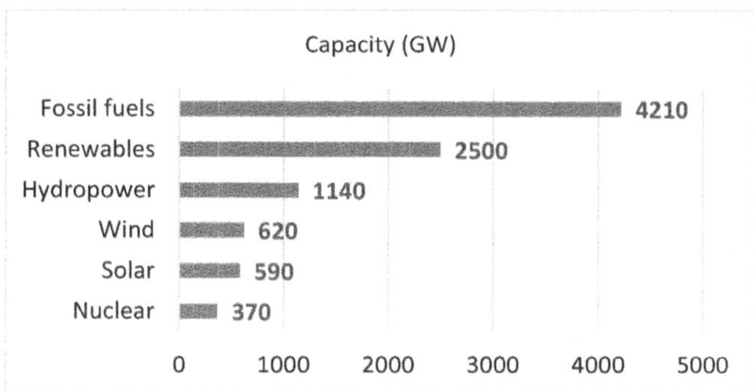

Figure 6.18 Installed electricity power capacity worldwide in 2019. Hydropower, wind, and solar are part of renewables and make up 94% of the renewables.

global withdrawals for energy generation roughly 11% are consumed (IEA, 2012). Roughly 50% total freshwater withdrawals are for cooling in Europe and the USA. The fraction of water use for power plant cooling is generally much lower in the low-income countries, where agriculture is the dominating water user.

The countries or regions with the largest withdrawal and consumption of water for energy production are the world's largest electric power producers, the USA, the European Union, China, and India. All of them have significant inland generating capacity to meet demand. On the contrary, countries such as Japan, Korea, and Australia have minimal freshwater requirements for energy because they can site virtually all of their power plants on the coasts and use seawater for cooling. Water scarcity is a major constraint on water use for energy production in the Middle East. Their power plants are adapted to scarce water conditions and have tried to minimize dependence on freshwater availability.

Cooling systems in thermal power plants are discussed in detail in another book[235], so here we present some key data to illustrate the interactions. Two broad categories of cooling system are available: once-through and re-circulating, which in turn is further divided into wet, dry, and hybrid systems. Each cooling system involves trade-offs in terms of water use, impacts on water quality, plant efficiency, and cost. The three most common cooling methods are:

> Important to distinguish between withdrawal and consumption.

- Open loop or once-through cooling systems – withdraws water, fresh or saline, for one-time use and returns nearly all the water to the source
- Closed loop systems – water is recirculated through the use of cooling towers
- Dry cooling systems – cools by use of fans.

Even if the abstraction is high, the water loss in open loop cooling is relatively small, some 3% of the water withdrawn. Closed loop cooling withdraws much less water than open loop cooling but consumes more water through evaporation. The closed loop cooling also requires more land area. Wet recirculating cooling systems are approximately 40% more expensive than once-through cooling systems. The order of magnitude of water withdrawal for once though cooling is depicted in Figure 6.19a, while consumption is shown in Figure 6.19b.

Carbon capture and storage (CCS) will be discussed in more detail below. Even if carbon capture technology is commercially available today there are currently only few large-scale commercial CCS power plants projects in operation. Part of the explanation is the high capital costs of the technology and the sustained operating costs. Another hurdle to reduce carbon emissions is the high additional water consumption, since additional electric power is needed to run auxiliary equipment, such as pumps, fans, and compressors for the CO_2 capture stream. This means that more fuel inputs are required to achieve the same electricity output, resulting in additional amounts of cooling water per kWh generated.

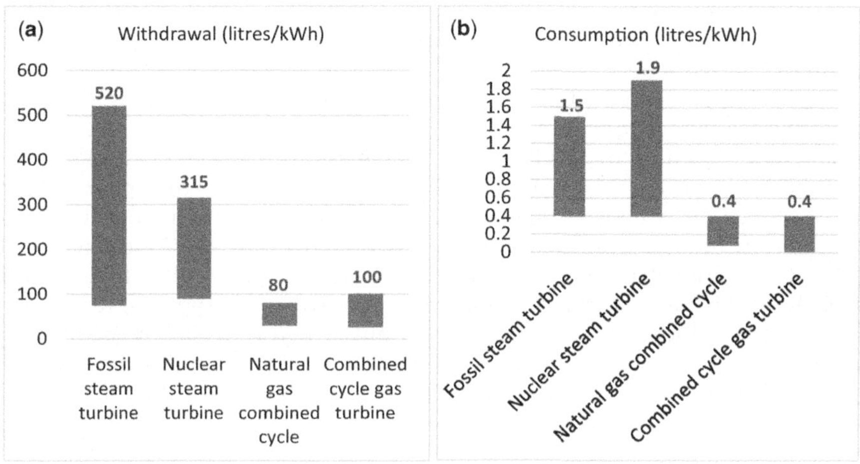

Figure 6.19 Water withdrawal in thermal electric power plants with (a) once through cooling systems (litres/kWh), and (b) magnitude of water consumption.

6.10 CARBON CAPTURE

Carbon capture, utilization, and storage (CCUS) is a common term for various technologies to capture CO_2 and use or store it to prevent its release into the atmosphere. This is also called 'negative emissions'. The captured CO_2 can sometimes be used to create products ranging from cement to synthetic fuels. CO_2 can be captured at the source of the emissions, such as power plants, or even from the air itself.

The basic method of CCS technology is to compress the CO_2 into liquid form once it has been captured. Then it is ready for transportation and storage. When it arrives via pipeline, truck, or ship the liquid is pumped into porous rock formations 1–2 km underground. The storage rock site has an impermeable layer (known as cap-rock) above the porous rock, which should prevent the liquid CO_2 from escaping. This storage mimics how oil and gas have been trapped underground for millions of years. Deep saline formations and depleted oil and gas reservoirs have the largest capacity. The first large-scale CCS project with dedicated CO_2 storage and monitoring was commissioned at the Sleipner offshore gas field in Norway in 1996. By 2020 the project has stored more than 20 Mt of CO_2, corresponding to about 0.05% of global *annual* emissions.

At current costs, fossil fuel-based power plants with CCS cannot compete with renewable power, as noted in section 6.9. For example, a combined cycle gas turbine plant (CCGT) with CCS has an LCOE that is 70–140% higher (including CO_2 transport and storage) than a plant without CCS. Furthermore, a higher power requirement will also require substantially more water for cooling.

An interesting argument favouring CCS comes from the oil industry. CO_2 can be injected into old non-profitable oil wells. In this way it is possible to enhance the oil recovery. The advocates suggest that the extracted oil will be carbon negative (!) since the CO_2 captured is more than what is contained in the oil that it extracts.

IRENA has presented a 1.5°C Scenario where 90% of the solutions to achieve net-zero emissions involve renewable energy. Carbon capture and storage may be part of the solution to handle industrial emissions, like cement and steel.

Direct air capture (DAC) captures CO_2 directly from the atmosphere, where its concentration is above 400 ppm (or 0.04%), which is high from a climate point of view but low from a carbon capture point of view. DAC facilities could be located at industrial locations. The air flows through a filter that absorbs the CO_2 by causing it to react with a potassium hydroxide (potash, KOH). The CO_2 can also adhere to a chemical membrane which traps it. The resulting is then mixed with calcium hydroxide (builder's lime). The lime absorbs dissolved CO_2, producing small flakes of limestone. The flakes are sieved off and heated in a calciner, until they decompose, releasing pure CO_2, which is captured and stored. The leftover chemical residues are recycled back in the process[236].

A DAC plant called Orka (Icelandic for energy) has been installed on Iceland in 2021 by the companies Climeworks AG and Carbfix. The CO_2 is dissolved in a liquid. The carbon-rich chemicals are then heated to about 100°C to release CO_2 as a pure gas that is pumped underground and injected into basaltic rock where the dissolved CO_2 crystallizes into solid carbonate minerals. The capacity is 4 Mt per year. Climeworks aims to get that cost down to US$200–300 a ton by 2030.

> Carbon capture
> is still too
> expensive to play
> a significant role.

Carbon Engineering's DAC in Canada is running a prototype and claims that the system can remove CO_2 from the air at a cost of US$100/ton[237]. Other sources estimate that the cost will approach the order of US$400 per ton of CO_2 in 2050. This is far too expensive to play any significant role in climate control. It looks far more efficient to spend funding on clean energy, like wind and solar, rather than building DAC systems.

Some simple calculations may give some perspective on DAC. The facility needs a solvent. To supply DAC plants to capture 10 Gt of CO_2 every year (about 25% of current emissions) will require around 4 Mt of KOH, more than the annual global supply today.

Captured carbon can also be reused (CCU) but is likely to be only a small part of carbon capture. Liquid CO_2 can be cured in concrete or used as a feedstock to make synthetic jet fuel. Another use is to manufacture sodium bicarbonate ($NaHCO_3$), also known as baking soda. This can then be used in a range of products, from eye drops to biscuits.

Economics is the challenge. Suppose that CCUS could increase from 50 million tons per year (Mtpa, or around 0.1% of global emissions) today to at least 500 Mtpa in 2030. Still, this would correspond to just a little more than 1% of today's emissions. For the most mature technologies the costs for capture, transport, and storage are in the range of US$22–225 per ton of CO_2. Capturing emissions from the smokestacks of factories or power plants before they escape into the atmosphere is much cheaper than DAC. The reason is that the concentration of CO_2 in these gases can be as high as 10%, rather than 0.04% in the air. The cost for CCUS is currently estimated to US$600–800 per ton CO_2 captured. Some theoretical studies not yet confirmed have estimates of US$94–232. Thus, the cost for 500 Mtpa would be of the order US$ 300–600 billion per year[238].

Carbon capture requires energy. The energy requirements differ by the technology used but are significant in all cases. It is estimated that around 200 TWh is required per 100 Mt of CO_2 captured. Thus, to capture 4 Gt by 2050 would consume 8000 TWh of electricity per year, which represents about a third of the global electricity use today. In IRENA's 1.5°C Scenario, electricity use increases approximately three-fold to reach 70 000 TWh, so the additional use would require a further 11%. That is an additional demand and comes on top of an already herculean scale-up in electricity supply. IEA presents a global overview of CCUS[239].

6.11 SOLAR AND WIND POWER

As shown in Figure 6.6 renewable power generation (excluding hydro) has surpassed nuclear power generation. This happened in 2020. The nuclear power generation has been almost flat since the year 2000. Only four more plants are in operation in 2021 compared to 2001.

There is a major difference in the structure of renewables compared to nuclear power. The renewables consist of more than 300 000 wind turbines, more than one billion photovoltaic (PV) modules and hundreds of geothermal plants over the world. Nuclear power, on the other hand, consist of much larger units and a much smaller number of nations. Nuclear technology today is dominated by boiling water and pressurized water reactors.

6.11.1 Energy efficiency and consumption

Wind and solar power will not be the only means to achieve emissions savings. Increased energy efficiency will be a crucial factor to reach net zero emissions. Filling leaks in home ventilation systems cuts the energy used for heating or cooling without sacrificing comfort. Achieving better fuel economy in petrol and diesel cars reduces fuel use but doesn't make the drive less pleasant. Switching from an internal-combustion engine car to an electric car means going the same distance for less than half the energy. Traveling by public transport instead of a car is a lot more energy efficient. Better insulation in buildings means less energy for heating and cooling. And so on.

Improving efficiency, like home insulation or investing in renewable energy like solar PV or wind, typically involves a high up-front cost. The return from lower or no fuel bills often takes years to materialize. The economically rational individual would make the investment knowing that long-term gains are worth it. However, only few act rationally, and many fewer have access to the capital to be able to act rationally.

There is a well-known problem that efficiency gains often lead to an increase in demand, known as the Jevons' Paradox, named after the 19th century British economist William Stanley Jevons:

- If the fuel efficiency in cars is improved, people tend to travel longer distances.
- By replacing traditional incandescent lamps with LED-lights the electricity bills have decreased, so people tend to leave the lights on for longer.
- The first T-Ford from the Ford assembly used around 10 litres of gasoline per 100 km. Many cars today use more gas. Surely, the engines are more efficient, but there is much more extra equipment like air conditioning, radio, and cameras.
- Aircrafts have been more than 70% more fuel efficient since the 1960s, but the total aviation emissions have increased about 4–5% per year.

As so often, Sir David Attenborough has formulated the task ahead so well: 'Tackling climate change is now as much a political and communications challenge as it is a scientific or technological one.'[240] That's especially true when it comes to energy efficiency. Of course, it does not hurt to become more energy efficient. However, predictions that new and efficient technology can save the climate should be taken with a pinch of salt. The consumption must decrease.

> Energy efficiency is good.
> Consuming less is better.

6.11.2 Solar

As shown in Figure 6.6 and in Table 6.2, there has been a phenomenal increase in the solar PV capacity in the world. The price reduction during the past decade has been superior, compared to all other electric power sources. The great advantage of solar PV is the scalability, from single households to large utility plants. A decentralized rooftop installation is truly decentralized and decreases the need for transmission capacity. On the other hand, it would require some local storage capacity, like batteries. Centralized solar PV production in utility solar farms will require more transmission capacity but decrease the need for local storage, since the grid can help to balance the load and the production. It is not a trivial task to evaluate the economics in the comparison between local storages and grid solutions.

> Huge areas are available for solar PV, our roofs. They can produce ~200 W/m^2 in good sunlight.

A standard solar panel has an input peak rate of about 1 kW/m^2. A modern solar cell has an efficiency of 20%, so the electricity production will be around 200 W/m^2 in good sunlight.

In a concentrated solar power (CSP) plant, solar thermal energy is used to make steam, which is thereafter converted into electricity by a turbine. There is a wide range of technologies to track the sun and to concentrate the sunlight.

A working fluid is heated by the concentrated sunlight and is used for power generation or as energy storage. This thermal storage makes it possible to generate electricity around the clock. For example, Morocco has built a huge CSP plant, called Noor (Arabic for light). In the middle is the tallest tower in Africa, 236 metres high. The plant occupies 3000 hectares in Ourarzazate at the edge of Sahara and has a capacity of 582 MW. Seven thousand mirrors reflect the sun rays towards the tower.

6.11.3 Wind power

Like solar PV, the growth and price development of wind has been impressive, as shown in Figure 6.6 and in Table 6.2. Wind power took off earlier than solar PV, but the latter is growing at a greater rate today. Offshore wind power can have a relatively high-capacity factor of around 50%, or 4000 hours/year.

Offshore wind turbines are getting huge. Soon a 15 MW wind turbine will be raised, which means that the maximum size of turbines has doubled in only 4 years. Even if there is a huge financial risk today, bigger turbines are more efficient in terms of installation time and cost. A 15 MW unit needs just one base structure and set of cables compared to two for a pair of turbines of half the size. The size of the turbine is gigantic. Each blade on a 15 MW unit is 115 metres and the rotor diameter is 236 metres. The tip of the blade has a maximum speed of 90 m/s or 324 km/h.

> Offshore wind power has a relatively high capacity factor.

The huge turbines are meant to be placed far away from land, which also means that the electrical transmissions must be longer. It is not evident if AC or DC is the best solution. With AC there are greater losses in the cables, while DC requires more advanced engineering. Floating 20 MW wind turbines are soon to be assembled and raised in harbours and sailed out to wind park locations for their 30 years offshore power generation duty. Limited maintenance is crucial.

Also, onshore turbines are getting huge, almost 280 metres tall, like the Eiffel Tower. This means they are visible far away and people will protest, even in sparsely populated areas. People like wind power but 'not in my backyard'. Views of natural beauty are disturbed, birds are threatened, and military interests are competing with energy demands. The resistance to wind typically magnifies in densely populated areas. Naturally, if wind power is located far away from populated areas – onshore or offshore – the transmission costs will increase.

The wind turbine output depends strongly on the wind speed and is related to the cube of the wind speed. In other words, if the wind speed decreases to half, then the wind power will decrease around 87%. Therefore, it is crucial to understand the wind patterns. The IPCC report suggests average European wind speeds will reduce by up to 10% because of climate change. However, near-surface wind speeds are extremely difficult to predict – much harder than surface temperatures. Some models show wind speeds increasing as temperatures warm, and others show decreases. There is much research going on to better understand wind patterns and how they relate to climate change.

6.12 ENERGY STORAGE

Electric power consumption is varying all the time, in timescales from seconds to seasons. Traditionally, electric power systems have controlled the production by frequency control. Hydropower is an excellent power source to adapt to consumption. In a longer timescale power plants can be started or shut down. Wind and solar create a new challenge, where both production and consumption are variable, seemingly randomly. It is obvious that intermittent production from wind and sun needs to be complemented with storage facilities.

> Storage technology should adapt to the time scale.

Battery storage is suitable for periods from seconds to days. For seasonal variations, for example to supply sufficient power in a cold winter day, there are other options like hydrogen gas and biogas turbines. However, traditionally electric power systems have been production oriented, simply aiming to produce what the users demand.

Demand management can be applied by influencing the customer behaviour:

- People can be discouraged from using electricity during peak hours and encouraged to use more off-peak electricity to even out the demand during the day.
- Peak and non-peak tariffs can be different.

Demand management can also be controlled by the utility. Today we have the technology to control individual loads remotely. By using a more flexible tariff structure, customers can be motivated to give various loads lower priority. For example, charging the electric car, starting the dishwasher, or using the washing machine does not need to take place immediately. Instead, the user can tell when the load must be completed, and the utility can turn on and off at suitable times. Naturally, demand management does not completely solve the intermittence problem but it will help.

Below we will consider two ways: batteries and hydrogen gas. The storage can be near the point of generation, at home, or within the grid.

6.12.1 Batteries

With intermittent production from solar and wind, batteries will play a key role for energy storage. Not all batteries are created equal, even batteries with the same chemistry. Batteries can be designed either to be high-power or high-energy and are often classified as one or other of these categories. We concentrate on two categories of batteries: lithium-ion (Li-ion) and saltwater.

Batteries are often evaluated according to their energy density (kWh/litre (dm^3) of battery) or total weight. This is important in electric vehicles and in many consumer products like mobile telephones. By this measure, lithium batteries have so far been the most successful. However, their cost is relatively high. A stationary battery can be assessed in a different way:

> Battery type and design depends on the application.

even if its energy density is less, the battery may still be successful; the total weight is not critical, so other properties can be considered. Consequently, a lot of attention is directed towards flow batteries and saltwater batteries. Typically, a Li-ion battery has an energy density of 100–260 Wh/kg, while a vanadium-based flow cell battery (for stationary storage) has around 20 Wh/kg, but 2–3 times higher seems to be reachable. As a remark, gasoline has about 150 times higher density than a Li-ion battery. This is of course one of the secrets behind the success of fossil fuels.

A lot of research is going on to find ways to increase the energy density in batteries. One example is alkali metal–chlorine batteries that could store six times the charge of existing Li-ion batteries. Development is still in the early stages. For now, the target is small consumer electronics.

6.12.1.1 Batteries for transportation

The battery market is dominated today by Li-ion batteries and demand for these batteries has created a large demand for critical minerals like lithium, nickel, and cobalt. They were initially developed through public and private research that took place mostly in Japan. Their first energy-related commercial use was developed in the USA, while most of the manufacturing today is in China. Li-ion batteries can be charged, discharged, and recharged hundreds of times before expiring. Compared to other common types of batteries, Li-ion batteries also tend to have a higher energy density, voltage capacity, and lower self-discharge rate too. This makes them most attractive for compact energy storage, like in phones and in electric vehicles.

There is a lot of research on various kinds of batteries, based on other components:

- Iron and oxygen
- Liquid calcium alloy and molten sodium
- Molten sodium and sulfur.

Still these designs are expensive and/or work with high temperatures. Researchers are positive that some of these designs can be much cheaper than Li-ion batteries within 10 years. This would be a formidable progress for renewable energy.

6.12.1.2 Stationary energy storage – flow cell batteries

Stationary batteries should be evaluated differently from batteries for transportation. A flow battery, also called a redox flow battery, is an electrochemical cell. Two chemical components dissolved in liquids are contained in the system and separated by a membrane. The chemicals provide chemical energy that is converted to electrical energy via the membrane. There is ion exchange through the membrane, while both liquids circulate in their own respective volumes. The ion exchange is followed by an electric current.

The fundamental difference between conventional batteries and flow batteries is that while energy is stored as the electrode material in conventional batteries, it is stored in the electrolyte in flow batteries. A wide range of chemistries and electrolytes have been tried for flow batteries, such as zinc–bromine, all-iron, semi-solid lithium, and hydrogen–bromine. Vanadium is increasingly being embraced by battery manufacturers as a core material in the production of batteries to be used in both small-scale and large-scale applications. It was invented by Professor Maria Skyllas-Kazacos at the University of New South Wales, Australia in 1983[241]. The electrolyte is composed of vanadium salts (in the range 1–2 mol/dm^3) in sulfuric acid (around 2–4 mol/dm^3). The vanadium ions serve as charge carriers. The reason that vanadium is so interesting is its ability to exist in solution in four different oxidation states. As a result, the battery needs to use only one electroactive element instead of two. This means that both the electrolytes on each side of the membrane are vanadium-based and there is no risk of cross-contamination.

> Electrolysis technology becomes increasingly important.

Vanadium redox-flow batteries (VFBs) have several interesting properties:

- There is no limit on the energy capacity. The battery has an extremely large overload capacity.
- The battery can remain discharged indefinitely without damage.
- The electrolyte is a safe and non-flammable liquid. The electrolyte solutions contain more than 50% water.
- The operating temperature has a wide span.
- The lifetime is long, at least 15 000–20 000 cycles.
- The electrolyte can be reused *ad infinitum.*

As mentioned, the energy density is relatively low and the weight is high, but this is not a major disadvantage for a stationary battery. Also, oxides of vanadium are relatively toxic.

The battery can serve as a rechargeable battery, where an electric power source drives regeneration of the fuel. One of the biggest advantages of flow batteries is that they can be almost instantly recharged by replacing the electrolyte liquid. The energy capacity is a function of the amount of liquid electrolyte and the power a function of the surface area of the electrodes. During a VFB charge, V^{3+} ions are converted to V^{2+} ions at the negative electrode through the acceptance of electrons. Meanwhile, at the positive electrode, V^{4+} ions are converted to V^{5+} ions through the release of electrons. Both reactions absorb the electrical energy put into the system and store it chemically. During discharge, the reactions run in the opposite direction, resulting in the release of the chemical energy as electrical energy.

Typical commercial module sizes are 1–1.2 MW. Activities in flow batteries in China illustrate the growing interest in using flow batteries for energy storage. Two projects in Hubei and Dalian, Northern China, are sized at 100 MW/500 MWh and 200 MW/800 MWh respectively. The Chinese government is aiming for a long-term strategy to push energy storage to integrate renewable energy. The system installation cost (CAPEX in US$/kWh) for flow batteries is only 15% higher than Li-ion batteries in grid scale today. Within 4–5 years it is predicted that the flow batteries will be marginally more expensive than Li-ion batteries.

There are three key producers of vanadium today, located in Brazil and South Africa. They produce around 20% of the global vanadium supply by mining. Around 70% comes from steel production, where vanadium is a by-product. The rest, 10%, comes from recycling of spent oil-refining catalysts that contain vanadium. Only less than 2% of vanadium demand comes from flow batteries, so at the moment vanadium availability does not seem to be limiting flow battery demand.

6.12.2 Hydrogen

Hydrogen is considered an attractive energy storage material both for long-term energy storage in intermittent power supply and for the energy source in electric cars. Hydrogen is an efficient energy carrier with a high energy density, considerably more than Li-ion batteries, gasoline, and ethanol (Figure 6.20). Obviously, hydrogen gas is the lightest gas, so to use hydrogen for energy storage it is typically liquified. To obtain hydrogen as liquid at atmospheric pressure it is cooled down to $-253°C$ (or 20 K). The liquid hydrogen

> H_2 has a high specific energy per kg. Volumetric energy comparable to gasoline.

density is low compared to other common fuels, 71 g/litre or just 7% of water density. Still, it is 800 times denser than hydrogen gas. Consequently, even if the specific energy per kg is high, liquid hydrogen has a volumetric energy density comparable to that of other fuels (Figure 6.20). To store liquid hydrogen requires cryogenic technology with special thermally insulated containers. However, it is still difficult to keep such a low temperature, and the hydrogen will gradually leak away. Also, all forms of hydrogen must be handled with competent safety.

6.12.2.1 Hydrocarbon route

Some 95% of the hydrogen used today is made from fossil fuels via the hydrocarbon route. This is a thermochemical process using heat and chemical reactions to obtain hydrogen from organic materials like fossil fuels. This is the

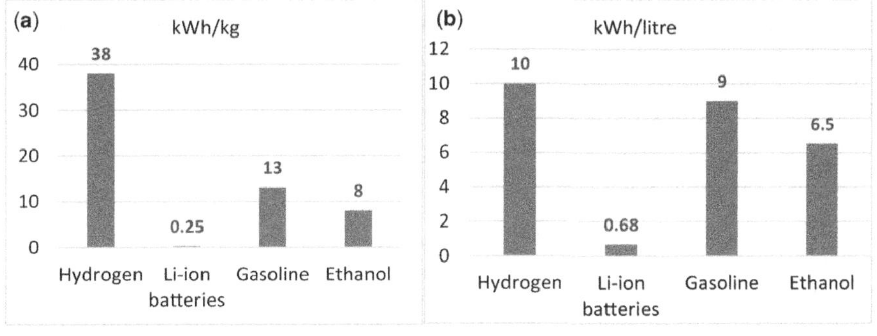

Figure 6.20 Energy density of common energy carriers, expressed as kWh/kg (a), and kWh/litre (dm³) (b), respectively.

most economical production today, but the cost is the carbon emission. The hydrogen source is mostly natural gas. High-temperature steam (700°C–1000°C) under high pressure, some 3–25 bar (0.3–2.5 MPa), reacts with methane in the presence of a catalyst:

$$CH_4 + H_2O \ (+\ heat) \rightarrow CO + 3H_2$$

Using carbon capture and storage will decrease the carbon footprint but as noted in section 6.10, there are several hurdles that must be overcome.

6.12.2.2 Electrolytic production

Water can also be split into hydrogen and oxygen by supplying electric power into an electrolysis process, since water cannot be split into its components spontaneously. An electrolyser unit is scalable from small-scale units to large central production facilities. An electrolyser consists of an anode and a cathode separated by an electrolyte. Different electrolysers function in different ways, mainly due to the different type of electrolyte material involved and the ionic species it conducts. Here we mention one kind to describe the idea of hydrogen production using a polymer electrolyte membrane electrolyser[242]. The electrolyte is a solid specialty plastic material. In acidic conditions, water reacts at the anode to form oxygen and protons (H^+):

$$2H_2O \rightarrow O_2 + 4H^+ + 4e^-$$

The protons move across a membrane in the electrolyte towards the cathode, while the electrons flow through an external circuit. At the cathode, the protons combine with electrons from the external circuit to form hydrogen gas:

$$4H^+ + 4e^- \rightarrow 2H_2$$

Since hydrogen is supposed to be part of green technology, it does not make sense to provide the necessary electrical power to the electrolysis via fossil fuels. Instead, renewables (wind, solar, hydro, geothermal) and nuclear energy options are considered to produce what is called *green hydrogen*.

To develop the green hydrogen technology is a key priority in many countries. Today only around 5% of industrial hydrogen is produced by electrolysis as a by-product from chlorine production. A detailed overview of green hydrogen production and its technology and economy has been presented by IRENA[243]. The cost of hydrogen through electrolysis would fall if a cheap source of electricity could be found. Also, water availability is of course crucial, since roughly 9 m³ of purified H_2O is required for each tonne of H_2.

6.12.2.3 Photosynthetic hydrogen

To produce molecular hydrogen using photosynthetic microorganisms has a lot of promise. This formation takes place under normal ambient temperatures. The hydrogen is formed using macro- and micronutrients, water, and sunlight. The great promise is that no greenhouse gas is formed. However, still this kind of production has low yield and efficiency and today it cannot compete economically with hydrogen production using fossil fuels.

6.13 CRITICAL RESOURCES FOR RENEWABLE ENERGY

Oil and gas have been associated with geopolitics. One of the great advantages of solar and wind is that the power can be generated anywhere. However, critical minerals needed for energy storage are becoming a geopolitical issue. Growth in battery demand translates into an increasing demand for critical minerals, and they are concentrated to a few countries. This may quickly become a source of concern because of trade restrictions or political instability. The metals needed in making rechargeable batteries used in everything from electric cars to energy storage to smart phones, include cobalt, lithium, and nickel. As expressed by Dr Fatih Birol, the IEA Executive Director: *'Today, the data shows a looming mismatch between the world's strengthened climate ambitions and the availability of critical minerals that are essential to realising those ambitions.'*

> Today there is a mismatch between renewable energy and extraction of critical minerals.

Critical mineral use come with land use changes and has a significant impact on water resources and water quality. There are alarming reports on corruption, human rights abuses, and the use of child labour in connection with critical mineral extraction. As reported by UNCTAD (UN Conference on Trade and Development)[244] and IEA[245], mining operations and mining companies have a huge role in the extraction of critical metals. IEA predicts that the global market for critical metals will approach what the coal market is today, and coal will decrease correspondingly[246].

Renewable energy plants have a different demand of critical raw materials compared to conventional power production. Solar PV and wind farms do not depend on fuel for their operation. However, they generally require more materials than traditional fuel-based counterparts for construction. Consequently, replacing fuel-intensive systems with solar and wind implies that fuel-intensive systems are replaced by material-intensive systems. The rapidly increasing demand of these resources makes it crucial to develop the circular economy where recycling of batteries, solar cells, and wind turbines will be a natural ingredient. However, both technology and recycling capacity are lacking today.

A typical electric car requires six times the mineral inputs of a conventional car and an onshore wind plant requires nine times more mineral resources (measured as kg/MW) than a gas-fired plant of the same capacity[247]. Since 2010 the average amount of minerals needed for a new unit of power generation capacity has increased by 50% as renewables increase their share of total capacity additions. Considering this development, it should be noted that during the period 2000–2018 the mining output in Europe has *decreased* by 19% while it has increased more than 100% in both Asia and in Oceania. In North and South America and in Africa mining operations have increased more than 20%.

Mining operations are always connected to various environmental consequences, at least from three aspects:

- *Land use*: This has a direct impact on people, biodiversity, and ecosystems. Often it will result in the displacement of communities.
- *Water use*: Mining generally requires large amount of water for the operations. Most often the produced and polluted water is not treated properly.
- *Water quality*: Mining operations are often a source of serious contamination, by acid mine drainage or wastewater disposal of tailings.
- *Solid waste*: Many of the residues of mineral development are hazardous to human health. They may also cause air pollution from particulate matter (e.g., mine dust) and gaseous emissions.

6.13.1 Cobalt

The mineral heterogenite can be refined into cobalt, one of the elements used in Li-ion batteries. Among other things, cobalt keeps the batteries, which power everything from cell phones to electric cars, from catching fire. Cobalt is also used in the manufacture of corrosion-resistant alloys for aircraft engines and gas turbines.

In order to illustrate the dark sides of cobalt extraction we will cite detailed reports from the city of Kolwezi and surrounding villages, located in southern province Katanga in the Democratic Republic of the Congo. The reports come from *The New Yorker*[248], Sky News[249], and Amnesty International[250].

Half the cobalt reserves in the world is found in Katanga, some 3.4 Mt, and in 2020 more than 67% of the world production came from Congo. The number two producer is Russia with 5%. Still, according to the World Bank, in 2018 three out of four people in Congo lived on less than 2 US$/day. More than 85% of the Congolese work informally, in precarious jobs that pay little, while the cost of living is remarkably high. Infrastructure has been ravaged by decades of dictatorship, civil war, and corruption. Food and other basic goods are often imported since there is little agriculture. The lure of mineral riches in a country as poor as Congo provides irresistible temptation for politicians and officials to steal and cheat.

6.13.1.1 The human cost

Many Congolese are employed at industrial mines, while others are 'diggers' or *creuseurs*. They work freelance at officially licensed pits, but many more hide and dig at night risking cave-ins and other dangers. It is relatively easy to extract cobalt. The minerals are relatively close to the surface, and they could be mined without digging deep holes. However, the *creuseurs* that had to dig deeper were digging more than 30 metres deep. Mine shafts have collapsed, killing miners. Still, people kept digging.

> Some 20% of the cobalt currently exported from Congo comes from artisanal miners in Katanga.

It is estimated that around 20% of the cobalt currently exported from Congo comes from artisanal miners in Katanga. The 2016 Amnesty International report claims that there are approximately 110 000–150 000 artisanal miners in this region, who work alongside much larger industrial operations. Of the 255 000 Congolese mining for cobalt, 40 000 are children, some as young as 6 years[251].

Congo and in particular Katanga has a violent history because of its wealth. In 1885, Belgium's King Leopold II claimed the country as his private property and brutally exploited it for rubber and millions of Congolese were killed[252]. Large-scale commercial mining began in Katanga in the 20th century. Artisanal mining became a source of livelihood for many people when the largest state-owned mining company collapsed in the 1990s.

The human cost for cobalt digging is terrifying. Teenage boys often work unprotected too near large mines. The prostitution of women and young girls is pervasive. Other women wash raw mining material, which is often full of toxic metals. The first study of the effects of mining-related pollution on new-borns in sub-Saharan Africa was made between 2013 and 2015 in the mining town of Lubumbashi in Katanga[253]. Women in Lubumbashi 'had metal concentrations that are among the highest ever reported for pregnant women.' The study also found a strong link between fathers who worked with mining chemicals and foetal abnormalities in their children, noting that 'paternal occupational mining exposure was the factor most strongly associated with birth defects.' A local doctor explained that many of the babies he delivered had mysterious illnesses: 'There are lots of infections they're born with, sometimes rashes, sometimes their bodies are covered in spots'. Chronic exposure to dust containing cobalt can result in a potentially fatal lung disease, called 'hard metal lung disease.' Skin contact with cobalt can lead to dermatitis (skin problems). The reports have found that a majority of miners do not have the most basic of protective equipment.

Researchers estimate that thousands of children work in mining in Kolwezi alone. The children are often drugged, to suppress hunger, according to Sister Catherine Mutindi, the founder of Good Shepherd Kolwezi, a Catholic charity: 'If the kids don't make enough money, they have no food for the whole day. Some children we interviewed did not remember the last time they had a meal.'

Artisanal mining supports poor families in the region, but the *creuseurs* suffer a lot and many of them have a short life. Many have physical and psychological injuries from mine collapses and other accidents, and from violent confrontations with the police and the Army. There are thousands of unofficial, unregulated, unmonitored mines where men, women, and children work in what can only be described as horrendous conditions. In one group, reporters found a circle of children with a four-year-old girl picking out cobalt stones.

In December 2019, attorneys from International Rights Advocates, a law firm in Washington, DC, sued Apple, Google, Dell, Microsoft, and Tesla for involvement in the injuries or deaths of child miners. *'These boys are working under Stone Age conditions for paltry wages, and at immense personal risk, to provide cobalt,'* the complaint alleges.

6.13.1.2 Environmental impact

Cobalt is a crucial component in global greening. Still its large-scale extraction generates large amounts of CO_2 and nitrogen dioxide (NO_2) emissions together with a lot of electrical power consumption for the operations.

The impact on water quality is huge. Cobalt mining sites also often contain sulfur-containing materials that can generate sulfuric acid (H_2SO_4) when exposed to air and water. When this acid drains from the mines, it can devastate the surrounding waters for a long time.

Streams alongside the main roads teem with women and children washing, sifting, and sorting minerals close to the mines. Hundreds of women with babies and infants work on the shore of Lake Malo, outside Kolwezi, to wash cobalt ore. Women suffer frequent urinary tract infections, which they attributed to working in dirty water all day. These infections could be caused by bacteria in the water, due to people defecating nearby. People must drink water from the mining sites after all the minerals have been washed in it.

> Producing critical metals for renewables has a huge impact on water resources and water quality.

In another village some foreign businessmen opened a cobalt-processing plant. Traditionally the villagers had farmed the surrounding bush, growing large crops of manioc. Due to the processing plant the land became polluted and left no source of employment for the villagers, except as low-paid day labourers.

6.13.1.3 Ownership

Huge sums of money change hands between large western and Chinese companies in the region. Today most of the cobalt in southern Congo comes from industrial mines, which are largely owned by Chinese companies. The mineral rights in the region were acquired in 2015 by a subsidiary of Zhejiang Huayou, a Chinese conglomerate. Huayou has made a huge investment in Congo and in 2017 they controlled 21% of the global cobalt market. There is little regulation requiring companies to trace their cobalt supply lines. As expressed by a Chinese financier working with mining deals in Congo: 'there is corruption, there is lack of the rule of law, which gives you more autonomy to be entrepreneurial.'

6.13.1.4 The supply chain

Amnesty International and others have reported on the supply chain of cobalt. They found that no country legally requires firms to publicly report their cobalt supply chains – allowing multinationals easy deniability.

Cobalt ore from *creuseurs* in Kolwezi are traded at a market in nearby Musompo. Independent traders, mostly Chinese, buy the ore and usually do not ask from where it has come. They sell the ore to larger companies, such as Huayou, which process and export it, mostly to China. There the cobalt is further processed and sold mostly to battery component manufacturers in China and South Korea. Battery manufacturers buy the materials and then sell the batteries to well-known electronic, smart phone, or electric vehicle brands.

6.13.1.5 Support for the miners

Some global big-tech companies launched the project *Cobalt for Development* in 2019, responding to the pressures[254]. With this project BMW Group, BASF, Samsung SDI, and Samsung Electronics, and Volkswagen are aiming to support ethical and safer practices in the DRC's cobalt mining industry. Another initiative is the Fair Cobalt Alliance, led by Tesla Inc., also backed by Zhejiang Huayou Cobalt Co. The projects will analyse how the workers' lives, work environment, and communities can be improved. This initiative includes efforts to increase education and arranging workshops on topics ranging from bread-making to women's rights, positive parenting, and conflict resolution. There is a hope to diversify incomes and reduce or eliminate families' reliance on child labour.

6.13.2 Lithium

The top lithium producing nations in 2020 were Australia (40 000 tons), Chile (18 000 tons), China (14 000 tons), and Argentina (6200 tons)[255]. The four nations produce 86% of the world production. Since 2016 the global lithium production has more than doubled. US companies have invested heavily in lithium mining projects in Chile, Australia, and Argentina; the production within the USA is only 1% of the world's lithium. China now controls about two-thirds of the lithium cell production industry.

Figure 6.21 shows the where the major lithium reserves are found. In South America lithium is found in vast plains of salt. The Atacama Desert in Chile has one of the world's largest lithium deposits. The areas have a volcanic history. Massive lakes were formed after eruptions and salt lakes were developed via evaporation. Lithium extraction has not only environmental impact. In Salinas Grandes, Argentina, indigenous people see natural pools as 'eyes' with spiritual meaning[256]. The biggest deposit of lithium is probably located in Bolivia, the Salar de Uyuni salt flat.

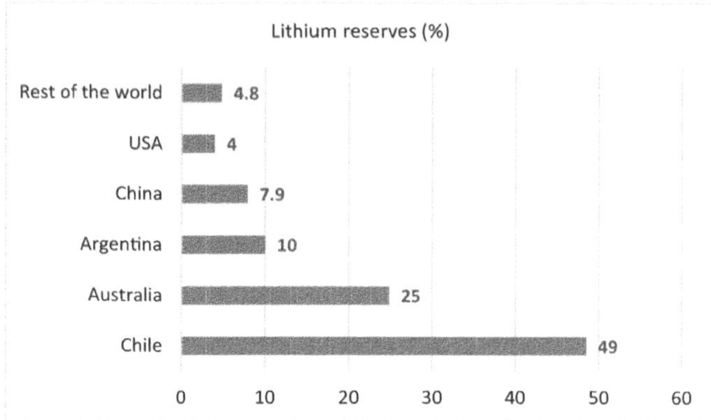

Figure 6.21 The global distribution of the largest lithium reserves. (Source: BP, 2021).

A hole is drilled in the salt flat and salty brine containing lithium is pumped to the surface. The water will evaporate, but this process may take several months. The resulting mixture of lithium, manganese, potassium, and borax (sodium borate) salts are filtered and placed into another evaporation pool. After another 1–1.5 years, the mixture is filtered sufficiently that lithium carbonate can be extracted. The large evaporation pools are often far from sealed. This can, and has, led to the leaching of toxic substances into the surrounding water supply.

6.13.2.1 Water impact

There is a huge water demand to extract lithium. Lithium is particularly vulnerable to water stress, since more than 50% of today's lithium production is in regions with high water stress levels. The estimate of the water requirement varies, but for every kg of lithium it is required 1000–2300 litres of water. The Atacama Salt Flats in Chile is one of the driest places on earth. Mining activities consume some 65% of the region's water, which is having a large impact on local farmers. The water consumption from lithium extraction has been connected to declining vegetation, increasing droughts in national reserve areas, and hotter temperatures. In Chile, the landscape is marred by mountains of discarded salt and canals filled with contaminated water with an unnatural blue hue. It is a troubling reminder that electric vehicles can help to reduce CO_2 emissions, but the battery production has a significant environmental impact.

> Producing 1 kg of Li requires 1–2.3 m^3 of water

More frequent droughts due to climate change has worsened the water stress. For example, in 2019 the operations in Chile experienced the most severe drought in 60 years.

Other lithium extraction technologies are getting developed where evaporation of water is replaced by chemical removal of the impurities. In this way the lithium is extracted directly from an unconcentrated brine to produce a lithium eluate, without using evaporation ponds.

6.13.2.2 Other lithium sources

Lithium is also found in rocks. Hard lithium mining requires drilling equipment to extract the lithium ore, which provides better quality lithium but at a higher cost. In Australia and North America, lithium is mined from rock using chemicals to extract it into a useful form. In Nevada, researchers found impacts on fish more than 200 km downstream from a lithium-processing operation. In Germany, lithium is mined from zinnwaldite, a silicate mineral. In the Eri Mine Works in Turkey lithium is extracted from the waste of boron production.

Even lithium extraction has geopolitical aspects. China's Tianqi Lithium owns a majority share of the huge lithium reserve in Australia, giving it a controlling interest. In 2018, the company became the second-largest shareholder in Sociedad Química y Minera – the largest lithium producer in Chile. China is by far the largest producer of lithium batteries.

It is estimated that 10 Mt of lithium, cobalt, nickel, and manganese will be mined worldwide in this decade for new batteries. Quite naturally there is a lot of research and development on battery recycling. Industry analysts predict that by 2020, China alone will generate some 500 000 metric tons of used Li-ion batteries and that by 2030, the worldwide number will hit 2 million metric tons per year. This has led and will continue to lead to large amounts of spent batteries ending their days in landfill. This is incredibly wasteful, as many of the main components, like lithium, could be recovered and reused.

6.14 ENERGY AND PUBLIC HEALTH

Air pollution is the single greatest environmental risk factor for premature death globally. Poor air quality is causing a range of diseases, from cardiovascular diseases (affecting the heart or blood vessels, like smoking) to respiratory diseases (the whole spectrum from common cold to asthma and lung cancer). The ultimate consequence of poor air quality is too-early death.

Particulate matter (PM) is the most dangerous component for human health and originates from almost all sectors such as power plants, industrial production, and road transportation. Typically, air quality is measured with the concentration of PM2.5, particles with a diameter less than 2.5 micrometres. They can penetrate the lung barrier and enter a person's blood system. The second cause of disease and death is ground-level ozone.

> Outdoor and indoor air pollution are the greatest killers.

The World Health Organization (WHO) estimates that 8 million people die prematurely every year due to air pollution, involving a combination of indoor and outdoor air pollution. All of this is man-made. To get it into some perspective: every year some 75 000 people die due to war and terrorism (statistics before the war in Ukraine 2022), and 1.3 million people die in road accidents, as illustrated in Figure 6.22. Also in Europe air pollution is a serious problem[257]. In 2018 alone some 417 000 people died prematurely due to PM2.5 in the air.

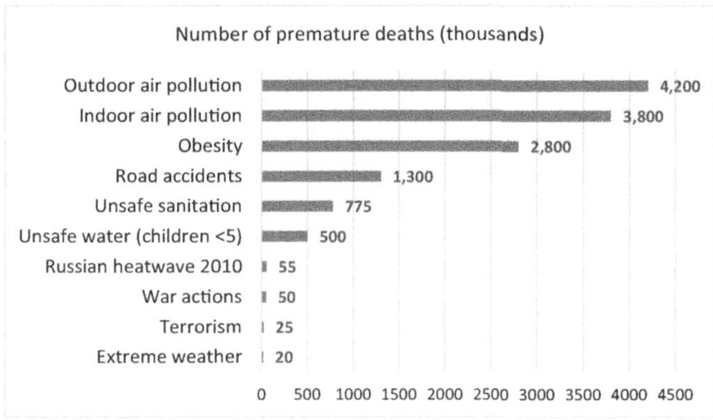

Figure 6.22 Causes of deaths during 2021. Most of the deaths are premature, caused by pollution, poor health, or unsafe water and sanitation[258].

The small PM2.5 particles typically are generated from combustion of fossil fuels, mostly from electricity and heat production, road transportation, and from industry. A significant fraction is also coming from residential cooking and heating (using wood and coal burning) in poor regions. Agriculture is the dominating source for ammonia while heavy metals are generated from heat and electric power generation, industrial production, and road transportation. PM2.5 particles typically contain:

- Sulfates (SO_4^{2-}) and sulfur dioxide (SO_2)
- Nitrogen oxides (NO_x)
- Ammonia (NH_3)
- Sodium chloride (NaCl)
- Volatile organic carbons
- Black carbon,
- Mineral dust and heavy metals, including lead (Pb), cadmium (Cd), mercury (Hg), and nickel (Ni), that are suspended in the air.

Poor people often are doubly exposed, that is to both indoor and outdoor pollution. Indoor pollution also contributes to outdoor pollution outside the home. The WHO guideline value of air pollution is an annual maximum exposure of 10 μg/m^3. Many countries are currently far from this goal. In most low-to-middle income countries, much of the population is still exposed to pollution levels above 35 μg/m^3. For example, in India 90% of the population is subject to higher pollution levels than 35, and 99% breathe higher pollution levels than 25 μg/m^3.

It is estimated that natural (such as desert dust) and anthropogenic air pollution will shorten the mean life expectancy per affected person by about 26 years.

Emissions do not just have health-related consequences. Air pollution will cause acidification of soils and water resources, eutrophication of soils and water, and corrosion of buildings and materials.

The link between global warming and climate change owing to air pollution is becoming increasingly clearer. Air quality has a feedback influence on climate. The energy flux in the atmosphere is called radiative forcing, measured in watts/m^2. If Earth receives more incoming energy from sunlight than it radiates back to space, then the radiative forcing is *positive*, resulting in global warming. Greenhouse gases (CO_2, methane, nitrous oxide, and halogenated gases) contribute to positive radiative forcing. Aerosols like sulfur dioxide (SO_2) and nitrogen oxides (NO_x) contribute to a *negative* radiative forcing, indicating that Earth loses more energy to space than it receives from the sun, causing a cooling effect. Many aerosols are hydroscopic in nature and can form cloud condensation nuclei. This can increase cloud formation, and a cloud can cool or heat the atmosphere depending on light scattering or absorption.

> There is a clear link between global warming and air pollution.

6.14.1 Coal and public health

Globally, surface coal mining is causing a variety of public health problems, for example respiratory illness, cancer, cardiovascular disease, kidney disease, poor birth outcomes, and increased mortality[259]. Direct mechanistic links are not well understood, but epidemiological and environmental evidence from studies around the world points to the high likelihood of public health harm imposed, in particular by surface coal mining activities.

Power generation using coal creates more air pollution than other fuels, including nitrogen oxides (NO_x), sulfur dioxide (SO_2), particulate matter (PM), polycyclic aromatic hydrocarbons (PAH), and heavy metals. This contributes to public health problems, such as respiratory illness, cancer, cardiovascular disease, preterm delivery, and increasing mortality among both adults and children. PM emissions alone from coal-fired power plants are estimated to be responsible for as many as 52 000 premature deaths annually in the USA, 670 000 in China, and 80 000–115 000 premature annual deaths in India.

Coal is still a common fuel source for heating and cooking in individual households. As noted in Figure 6.22, indoor pollution is a most serious problem. Most households relying on burning solid fuels, including coal, are in China and in India. The health consequences are particularly high since the coal use occurs mostly among the poorest rural households. They often burn coal inefficiently and have poor ventilation, leading to significant pollution exposures, particularly among women and children.

6.14.2 Biomass burning

Burning of wood, landfill gas, and solids waste is also causing public health problems. One estimate projects 40 000 additional annual deaths in Europe caused by household or commercial power generation of these kinds of biomass[260].

6.15 ACTIONS NEEDED

Energy together with water is the foundation for our lives. Typically, we waste too much energy in wealthy countries, while poor countries have a desperate need for more energy to lift people from poverty.

The challenge is not only to find ways to replace fossil fuel-based energy with renewable energy. The demand side –consumer behaviour – must be changed. It is a combination of using technology to manage demand side and to be more efficient and frugal in using energy.

- Again, the obvious action is to stop extracting and burning fossil fuels as quickly as possible.
- Coal, gas, and oil extraction have huge environmental impact. Fossil fuels are a great threat to the earth even before they are burned.
- Air pollution caused by fossil fuel burning is killing people. Public health is another critical reason to phase out fossil fuels.

- As long as the absurd difference between energy consumption in wealthy and low-income countries is not recognized, there will be neither a solution of the climate crisis nor a sustainable future.
- Global car production is still of the order 90 million cars per year. As long as we have fossil fuel cars, they will emit greenhouse gases.
- The issue is not to replace all fossil cars with electric cars. This is not realistic. We must analyse the wider perspective: how to transport people and goods.
- Thermal power plants require huge volumes of water for cooling. Renewables like solar and wind offer a double gain, both less emissions and less water demand.
- The first generation of biofuel is not only inefficient; it is a serious competitor to food supply.
- Carbon capture will probably not play any large role to solve the emission problems.
- Energy storage is a critical part of the renewable energy development. However, considering the dependency on critical minerals and metals, it is crucial to develop recycling technologies for batteries.

doi: 10.2166/9781789062908_0137

Chapter 7
The food perspective

Drought and conflict are endemic in the Somali region. Climate change is causing less and unpredictable rains. Most of the pastoralists have lost their cattle and are surviving on the good will of relatives and tribal solidarity. Since 2020 the region has been suffering from a severe locust plague, ravaging agriculture and devouring almost anything in sight.

Most of the crops on the land have been lost or harvested too early in a desperate attempt to save it. The grasslands for the camels, cows, goats, and sheep are barren. People are on the brink of hunger and huge loss of cattle due to lack of food.

Permission from Oxfam International (www.oxfam.org) gratefully acknowledged. (Photo: Petterik Wiggers/Oxfam)

We are heading for a food crisis. This is not a new insight. The fear of population increase combined with rising hunger has been discussed since Thomas R. Malthus, the English cleric and economist (1766–1834), predicted in his *Essay on the Principle of Population* (1798) that lack of food would be the limit to population growth. Paul Ehrlich (see Prologue) argued that the population growth would lead to mass starvation.

Today, hunger is escalating again after decades of progress. FAO estimates that around 2.4 billion people did not have access to adequate food in 2020 and 930 million of them were severely food insecure.[261] Some 50% of them live in Asia, another 35% in Africa, and most of the rest in Latin America and the Caribbean. The Covid-19 pandemic may have made an additional almost 100 million people living in extreme poverty.

> 930 million people are severely food insecure

The reason for hunger is not always an inability to produce enough food, but is largely attributed to instability and socio-political and economic factors.

The UN agreed to define the sustainability goals in 2015 to reach zero world hunger (SDG2) in 2030. Then it was estimated that some 615 million people were hungry. Since then, there have been several disasters related to global warming, such as the plague of locusts in Kenya (section 4.3), spreading to Middle East and South Asia, cyclones bringing flash flooding to Zimbabwe and Mozambique. Now FAO estimates that there will still be 660 million people living in hunger in 2030. Perversely, it is poor farmers and their families – those who produce an important share of the world's food, and whose livelihoods depend on food and agriculture – that are among the most likely to experience hunger. Also, in high-income countries, there are too many people living in 'relative poverty' or 'food deserts' who cannot afford or access adequate, nutritious food. It is estimated that just in Europe and North America more than 17 million people could not afford a healthy diet in 2019.[262]

Food production is closely coupled to climate, water, energy, economy, and lifestyle, as indicated in Figure 7.1. Climate change has a profound impact on food production and food availability. Agriculture depends crucially on both water quantity and quality. Increasing meat consumption in the world has a huge impact on land use, water, and energy use for food production. The impact of food production is noted in most aspects of human life. Greenhouse gases are emitted from fertilizer production, agro machinery, transportation, and animal farming. Farming practices are always debated and have a huge impact on public health, water consumption, energy consumption, soil quality and, of course, productivity. Food waste is a huge global problem, and the impact on water quantity and quality are well recognized. The first generation of biofuels constitutes a major competitor between food and energy.

To understand how climate change can alter food production requires an understanding of how the changing agriculture sector affects socioeconomic conditions, from food prices to consumption patterns, and how these in turn affect food production. Only a thorough systems analysis can provide this kind of understanding. Wasted food is a crucial factor. If only 25% of the food

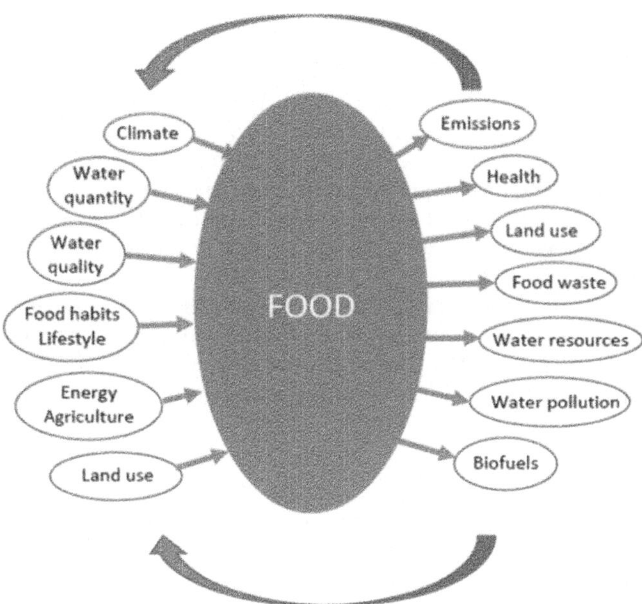

Figure 7.1 Interactions and feedback loops around food.

currently lost or wasted globally could be saved, it would be enough to feed nearly 900 million hungry people in the world.

7.1 FOOD AND CLIMATE CHANGE

Climate change is an agonizing threat to the food supply for the world. Year-to-year climate variability has a large influence on agriculture, which is heavily dependent in rainfall, sunshine, and temperature. The food crisis is significantly higher in countries where agriculture is highly sensitive to rainfall and temperature variability and extremes, and a large part of the population depends on farming for their livelihood.[263] At higher latitudes some producers may benefit from a longer growing season, but arid and semi-arid areas will experience increased water stress. Increasing frequency and intensity of extreme events, such as floods and droughts, will have an impact on crops and livestock, as discussed in section 4.3.

Climate extremes, like droughts, floods, hot spells or storms, have had particularly serious consequences in 133 low- and middle-income countries and have increased dramatically during the last 20 years[264]:

- In 2000–2004, 76% of these countries were exposed to climate extremes.
- In 2015–2020, this fraction had increased to 98%.
- 77% of small-scale farms across these countries are in regions with water scarcity. Only 37% have access to irrigation.

- 82% of all damage and loss caused by drought in 2008–2018 was experienced by agriculture.

Receding glaciers and rising sea levels and saltwater intrusion are great threats that have been mentioned in sections 3.2 and 5.2.

Global warming and higher CO_2 concentration will benefit certain species of weeds, insects, and other pests, increasing their ability to damage crops. The earlier onset of spring and warmer winters could allow some parasites and pathogens to survive more easily. In areas with increased rainfall, moisture-reliant pathogens could thrive.

Producing and transporting food generates greenhouse gas. Meat consumption is a huge contributor to both water use (section 5.6) and greenhouse gas emission. Food waste represents more than 20% of landfills and incinerators. Decomposition of food waste in landfills generates methane.

Greenhouse gas emissions related to food are substantial[265]:

> Food contributes with a quarter of all global GHG emissions

- Food contributes 26% of all global emissions.
- Animal products provide 58% of all food-related emissions.
- Beef and lamb make up 50% of all farmed animal emissions.

Many marine regions in the world suffer from overfishing and water pollution. Climate change may worsen these stresses. Temperature changes could lead to significant impacts and can affect the timing of reproduction and migration. The oceans are also gradually becoming more acidic since a large fraction of the atmospheric CO_2 is absorbed in the water. This could harm shellfish by weakening their shells, which are created by removing calcium from seawater.

Still food production is a great contributor to climate change with 26% of all greenhouse gas emissions, including contributions from land use (such as deforestation), agricultural production, processing, transport, distribution, and food waste.

7.1.1 Global meat production and consumption

In 1971 Frances Moore Lappé published her ground-breaking book on the environmental impact of meat production.[266] She argued that world hunger is not caused by a lack of food but by ineffective food policy. She noted that meat production is environmentally wasteful and contributes to the global food scarcity. She argued for a vegetarian lifestyle out of concerns over animal-based industries and animal-based products. The Lappé book was followed by another bestseller by Ellen Buchman Ewald in 1973, where she in a practical way explains how to cook responsibly for a small planet.[267]

Increased wealth has spurred people's appetites, boosting demand for luxury foods such as seafood and beef. The oceans are being emptied of fish: all but 10% of the large fish in the seas have been plundered. Rain forests are getting cleared not only for cattle but also for soybeans and oil palms planted not to grow food but to make biofuel to replace fossil fuels.

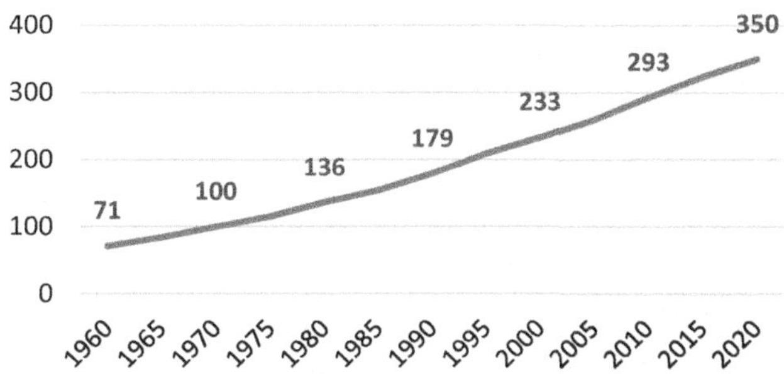

Figure 7.2 Global meat production (Mt) since 1961.[268]

Meat production has increased a factor of 5 since the early 1960s and it has doubled during the last 30 years, as noted in Figure 7.2. Production is expected to grow further. The most significant increase of meat consumption has taken place in China, from just 4 Mt in 1961 to 90 Mt today, while North America has increased in the same period from 17 to 47 Mt. So, today China consumes 26% of the global meat production, while North America uses 13%.

Meat production has increased almost twice as much as the global population. While meat production has increased 4.9 times since 1960 the global population has grown from 3.0 billion to 7.8 billion, a factor of 2.6. This corresponds to a world average of 45 kg/person. Most people will eat just a fraction of this.

The meat consumption in the most meat-eating and some of the least meat-eating countries is shown in Figure 7.3. Simply stated: rich people eat more meat. This is at least true when we make cross-country comparisons. The inequality is absurd.

If everyone would have the same meat-heavy diet as the average US person, then the world could feed only 2.5 billion people. Obviously, we cannot continue our current diet in rich societies.

Some numbers may illustrate the state of affairs:

- The world eats 40 000 tons of meat *every hour* – around the clock.
- Each kg of meat –on average – requires more than 6 litres of water.
- Each year, around 72 billion animals are slaughtered for human meat consumption, or more than 9 animals per human.

> Annual meat consumption = 9 animals per human in the world

- This includes 68 billion chicken (95%), 1.5 billion pigs, and more than 300 million cattle.
- The pork industry:
  - It is forecasted that it will continue to rise significantly.
  - It needs 30% of the world´s feed.
  - More than 50% of pork worldwide is produced and consumed in China. The country is already fighting against it by new dietary guidelines aimed to cut meat consumption of 50% by 2030.[270]

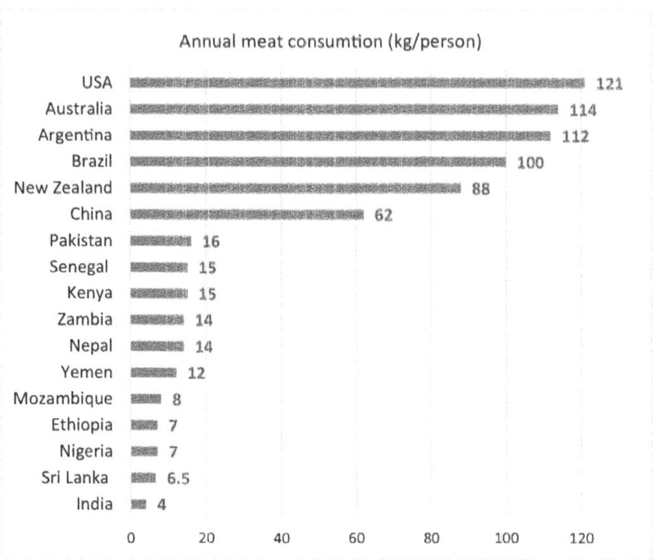

Figure 7.3 Average consumption of meat in some of the most meat-eating nations and some of the poorest nations.[269]

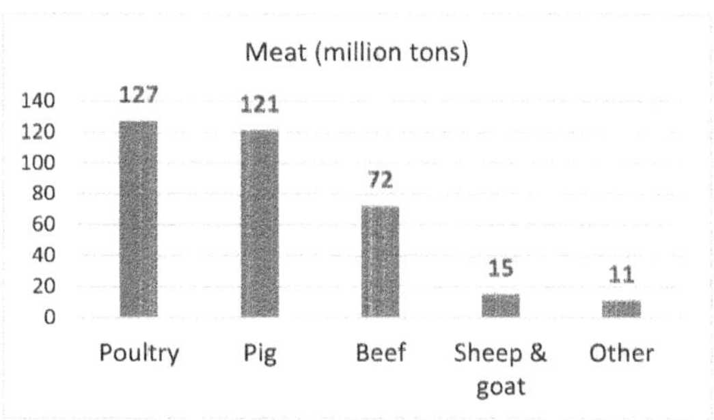

Figure 7.4 Types of meat produced annually in the world.

The size of global meat production is shown in Figure 7.4. The total annual global production is 350 Mt.

7.1.2 Environmental impact

The WWF has reported (2018) that 60% of global biodiversity loss is caused by meat-based products.[271] The current trend of meat consumption is increasing and contributes to most food-related greenhouse gas emission.

- It takes 75 times more energy to produce 1 kg of meat than 1 kg of corn.

- To produce feed for the cattle and other livestock animals in Europe requires a vegetation area seven times the size of EU.
- Most vulnerable regions like the Amazon, Cerrado (inner Brazil), the Congo Basin, Yangtze, Mekong, the Deccan Plateau (South India), and Himalayas are used as feed crop lands.

> 60% of global biodiversity loss is caused by meat-based consumption

Another way to express the climate impact:

- 1 kg of beef has a climate impact corresponding to around 9.5 litres of gasoline.
- Consuming 1 kg of beef per week corresponds to a yearly carbon footprint of almost 1400 kg CO_2. This corresponds to going by car some 8000 km, if the car uses some 6 litres of gasoline per 100 km.

> 1 kg of beef ⇔ almost 10 litres of gasoline

What will the huge animal food consumption mean for the health of animals, for people, and for the planet itself? The large amount of waste that cattle, pigs, and mass-produced poultry generate also has a large impact on water quality.

7.2 LAND AREA FOR FOOD

The global average of the total arable farmland has declined from 1.45 to 0.63 hectares/person from 1961 to 2018, a decrease of 57% (Figure 7.5). The cropland per person has declined 50%, from 0.36 hectares in 1961 to 0.18 hectares in 2018.[272] The world population increased from 3.07 billion in 1961 to 7.59 billion. The positive message is that while the population in 2018 was 2.47 times the population in 1961, the per capita arable land decrease was less than

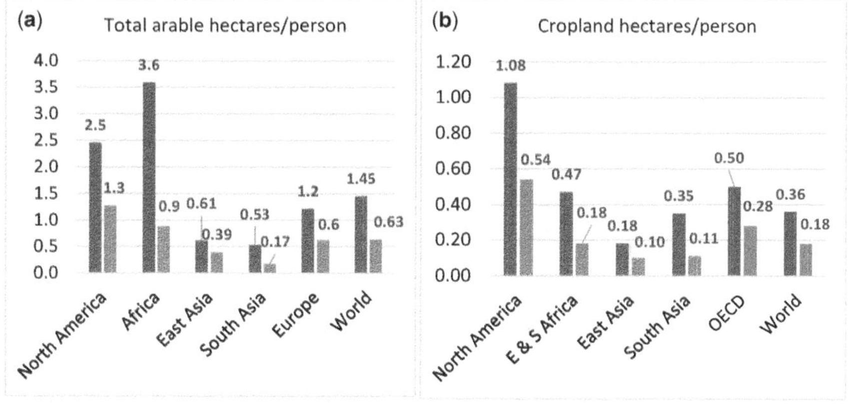

Figure 7.5 (a) *Total* arable land per person in some world regions (left bar = in 1961, right bar = in 2018). (b) *Cropland*: the most dramatic decrease has taken place in South Asia.

proportional. Naturally there are major differences of arable land in the world. Figure 7.5 summarizes arable land per person in 1961 and in 2018 across the world's regions. The available arable land in North America is greater than in all other regions, some 5–6 times larger than what is available in Asia. In South Asia the decline has been dramatic.

Farmland today occupies 38% of all land or 50% of all arable land, around 51 million km². This corresponds to the total land areas of the largest five countries in the world: Russia, Canada, China, USA, and Brazil. As noted in section 6.6, 40% of corn production in the USA goes towards ethanol and biodiesel. Most of the rest goes towards animal feed and a small fraction directly to the human diet, most as corn syrup, not so healthy food.

On a global scale 77% of all arable land (40 million km²) is used for meat and dairy production from livestock. This includes grazing land for the animals and arable land for animal feed production. The rest, 23% or 11 million km², is used for crop (excluding feed). Figure 7.6 illustrates the production of animal-based versus crop-based food. Despite its use of 77% of the arable land, meat and dairy supply only 18% of the global calorie production and 37% of the global protein supply. One explanation is also that livestock farming and cattle grazing can take place across a range of diverse climatic and environmental regions and is potentially less geographically constrained than arable farming.

> 3/4 of all arable land is used for animal food production

One way to describe the livestock is to describe the global mammal biomass. Livestock makes up 94% of the global mammal biomass, excluding humans. Wild mammals make up just 6%.

We need to increase our food production to feed a growing and increasingly affluent population. However, some farming methods are stripping nutrients from the soil, fuelling climate change, and driving biodiversity loss.

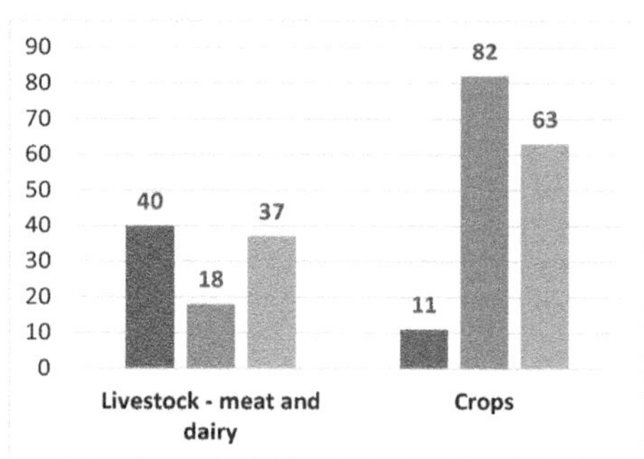

Figure 7.6 Land use for food production. The left bar shows the land use in million km². The middle bar is the percentage of the global calorie supply, while the right bar shows the percentage of the global protein supply.

Obviously, there is a strong interrelationship from food to water and energy use, to population and to national economies. Having less arable land makes it even more necessary not only to use water and energy more wisely but to reconsider agricultural practice. Increasing the production of food per area of land is of course crucial for food security, living standards, and for the impact on the environment.

There is a movement towards urban farming and with an increasing part of the global population in cities this ought to become an important complement to rural farms. However, to-date urban farms have mostly been limited to growing salads, herbs, and small fruits and vegetables. So far this provides a healthy diet but is still a low biomass production, but there are a lot of efforts to expand this food production in the vicinity of cities.

There is another challenge related to land use: who has access to the land. In many low-income regions, the producers are large landholders, and they may prioritize the cultivation of non-staple crops, such as tobacco, coffee, chocolate, peanuts, or exotic fruits for foreign markets. This will bring higher earnings but will harm the local farmer.

As described in section 3.7 large forests areas are cleared for local agriculture subsistence. The sad consequence is that the local farmer will have a smaller chance to provide food in the long run due to the loss of biodiversity.

7.3 FOOD WASTE

Quite often my wife and I discuss at breakfast: how far did our food have to travel before it reached our table? The supply chain of food is getting increasingly complex and sometimes absurd. Too much fruit and vegetables is imported, whereas it could be produced at home, at least part of the year. We can buy apples from our antipode New Zealand cheaper than apples from Sweden at harvest time at home and winter in New Zealand. Already today more than half the global population live in urban areas and an increasing part of all food will be eaten in cities. Long and complex food supply chains lead to waste. A few years ago, a Californian colleague, around 50 years old, visited us in Sweden. We went to a neighbouring farm. This was the first time my friend saw a living pig. In our time, the distance between the customer and the producer is sometimes huge.

There is enough food harvested for every person, around 4600 kilocalories per day. However, a large fraction is lost on the way. On average[273]:

- 14% of food is lost between harvest and distribution.
- In high-income regions, like North America and Europe, the loss is 16%.
- In lower income regions, the loss is less: around 12% in Latin America and 8% in Oceania.
- About 40% of all food is never eaten.
- 20–30% of all harvested grain gets lost during storage. The grain is sensitive to both temperature and humidity.
- 10% of the global population is hungry.

> About 40% of all food is never eaten

If food waste was a country, it would be the third highest emitter of greenhouse gases after the USA and China, according to FAO.

One reason for the greater loss in high-income countries is that we eat more (imported) fresh fruit, vegetables, and meat compared to low-income countries. If the food is not as high part of the total income, it will also contribute to more food waste. Consumers in high-income countries waste more than 220 Mt of food. The entire net food production in sub-Saharan Africa is 230 Mt. It is estimated that 1.3 billion tonnes of food are wasted every year before even reaching the shops. In other words: every hour some 140 tons of food is lost or wasted. This is around one-third of the food produced in the world for human consumption. High-income and low-income countries dissipate roughly the same quantities of food – each more than 600 million tonnes.[274] Almost *half* of all fruit and vegetables produced globally are wasted. In Europe and North America, we waste 95–115 kg of food every year, while people in sub-Saharan Africa, South and South-Eastern Asia, each throw away only 6–11 kg a year. We consumers must do our contribution to consume, not to waste food.

We know that a growing population will need more food and reducing waste is crucial. Furthermore, producing all the food that is never eaten requires huge resources in terms of labour, farmland, water, and energy. This has a great impact on global warming and loss of biodiversity which in turn increases food insecurity.

In low-income countries, the dominating causes – some 40% – for food waste is lack of cooling facilities and adequate transportation and storage before the food reaches the end user. In rich countries, market prices may govern the food waste. If they are too low, farmers may decide that it is too expensive to harvest. Many years ago, I witnessed this close to our home. Before the EU there were many Polish guest workers coming to Sweden (more or less legally) to work on the harvest. Close to our house there was a huge field growing cabbages. After one day of harvest, the Polish workers were expelled from Sweden. Swedish workers were not attracted by the low-paid job, so most of the harvest was left to rot in the field. Consumers often avoid products that are not perfect, so a lot of fruit, vegetables, and produce end up in solid waste or compost. In rich countries more than 40% of losses happen at retail and consumer levels.

Western Europe, the USA, and Canada have nearly twice as much food as is required by the nutritional needs of their populations. If crops fed to livestock are included, European countries have more than three times more food than we need. The USA has around four times more food than is needed and a large portion of the nutritional value is lost before it reaches people's mouths (see also Figures 7.2 & 7.3).

Today we are proud to recycle wasted food for composting. However, most uneaten food in high-income countries rots in landfills where it accounts for substantial methane emissions. For example, uneaten food in the USA contributes 25% of all methane emissions in the nation. Food manufacturing and consumption contain a lot of high-value fertilizers that

> In the USA: uneaten food contributes with 25% of all methane emissions in the nation

could be used to produce new food or the residues containing natural polymers can be used to manufacture other products. Wastewater treatment has been replaced by a better approach and attitude: 'water resource recovery' (section 5.3). Likewise, food waste is a resource.

7.4 WATER FOR FOOD

The water need in agriculture is huge, as described in sections 5.2 and 5.3 and the water footprint for various food products was illustrated in section 5.6. It is obvious that diets change as countries develop economically, when wealth increases, cheaper food becomes available, and global food markets are reachable. Livestock products require more water as measured by both weight and calories than cereals and other vegetarian food, as illustrated in Figures 5.10 and 5.11. The numbers shown are global estimates and often context-specific and cannot be generalized because of differences in feed used between and within species and production systems.

There is much less analysis of water use in seafood production. Fish is an important source of proteins, healthy fats, and nutrients. The water use in the seafood industry is very different, depending on if it is aquaculture or capture.

7.4.1 Irrigation, water efficiencies, and available technologies

Considering the huge water use in agriculture and food production, irrigation issues have a great impact on many activities. This was an issue already in the 1960s when Georg Borgström examined the impact of irrigation, as described in the Prologue. Figure 5.7 gave a quick overview of irrigation interactions. Here irrigation will be investigated from a food perspective: the water supply (this section), and the energy requirement (section 7.5).

Irrigation is particularly prevalent across South and East Asia and the Middle East. More than 50% of the agricultural areas in Pakistan, Bangladesh, and South Korea are irrigated. In India the rate of irrigation is 35%. The implication for groundwater resources were discussed in section 5.3. In Sub-Saharan Africa the rate of irrigation is lower.

In section 6.8 we commented on how water resources have a great impact on hydropower generation in water-scarce areas. Multipurpose reservoirs used for both irrigation and hydropower present obvious examples of the water–energy nexus. Water released for irrigation lowers reservoir levels and reduces hydropower generation, while hydroelectricity production may decrease water for irrigated food production. FAO[275] estimates that more than 500 GW of hydropower generation (out of global hydropower capacity of 1330 GW in 2020) competes with irrigation, for example in the USA, India, Central Asia, and Oceania.

> More than 500 GW of hydropower competes with irrigation

In secondary school I learnt that the Aral Sea was the fourth largest lake in the world. Today the lake is only one tenth of its size in the 1950s and looks like a depressing expanse of sand.[276] Because the Aral is/was an endorheic lake having no outflow, pollutants never washed out but sank to the bottom sediments. As the bottom became exposed to open air, salts and pesticides have been released into the atmosphere, leading to the destruction of farmland and causing public health problems, not only in the local region but thousands of kilometres away. The environmental catastrophe in the former Soviet Union of the disappearing Aral Sea may be the most frightening example of devastating misuse of water during the last century and a sign of human greed, hubris, ignorant politics, and lack of compassion.

Starting in the 1950s, water was heavily diverted from the Amu Darya and Syr Darya Rivers, flowing westwards from the Himalayas to the Aral Sea, to irrigate cotton, wheat, and other food grain farming. This was earlier an internal Soviet problem, but since 1991 it has become an international conflict that affects six countries. The Syr Darya in the north runs from Kyrgyzstan through Uzbekistan and Kazakhstan and the more southern located Amu Darya from Tajikistan through Uzbekistan and Turkmenistan. The Amu Darya and its tributaries form part of the border between the Central Asian states and Afghanistan. Water use has increased rapidly since the Central Asian states became independent and is now at an unsustainable level. The upstream country, Kyrgyzstan, controls most of the reservoirs regulating the river flow and stores water during summer for hydropower in winter, which then conflicts with the downstream countries' needs for irrigation during summer.

The Aral Sea catastrophe is also sad reminder of how societies and nations too often have tried to be the master of nature. Nature pays back. The huge mass of water in the previous Aral Sea regulated the regional climate. Today it is much harsher.

Unfortunately, there are other similar catastrophes. Lake Chad in West Africa used to be the world's sixth largest lake and has literally gone from being an oasis in the desert, to being just desert. Spanning the countries of Chad, Nigeria, Niger, and Cameroon and bordering the Sahara Desert, the Lake has contracted by a massive 95% between 1963 and 2001. Local fishers, farmers, and herders have lost their livelihoods. Deepening poverty has contributed to a collapse of law and order, growing jihadism, and an exodus of more than 2 million people, many heading for Europe. Climate change has been blamed for the drying up of the Lake. The initial decline in the 1970s and 1980s was due to long droughts. However, rainfall has recovered for two decades and still the Lake is quite empty. Rivers out of Cameroon, Chad, and Nigeria are being diverted to irrigate often inefficient rice farms instead of supplying water to Lake Chad. Hydrology analysis[277] found that water diversions for irrigation explained 73% of the reduction in flow into Lake Chad from the largest river, the Chari, since the 1960s. The diversion rose to 80% after 2000. Variability in rainfall explained just 20%.

7.4.2 Water quality

Agricultural production is becoming increasingly intensive, with high input of fertilizers and pesticides. The pollution load to the water environment through diffuse pollution is serious in many places. The fear of food scarcity is often in collision with environmental safety. Droughts and water scarcity create severe vulnerability of food supplies, food prices, and long-term sustainability.

As in most production systems, there is a trade-off between the desired outcome and the costs in terms of pollution and resource utilization. This is true for energy generation, and it is true for agriculture. Trade-off analysis has been used as a tool to evaluate the performance of food production.[278] To do this, we need reliable environmental and ecosystem models and great insight into food growth system behaviour and water quality management. However, the analysis must include the stakeholders linked to the land use and decision makers. In other words: smart use of nutrients and pesticides can boost yields while minimizing harm to the environment.

The development towards regenerative agriculture is described in 7.6.

7.5 ENERGY FOR FOOD

Energy is needed for food production, transportation, food processing, storage, and consumption.

7.5.1 Irrigation

Irrigation is an important part of agriculture, accounting for 20% of all farmlands and 40% of all production worldwide. In Figure 5.7 we summarized various impacts of irrigation. As noted in section 5.3, around 3 million km^2 are irrigated in the world and the water pumps consume more than 60 TWh/year (corresponding to about half the total Swedish electrical consumption). On top of this, energy is needed for the manufacture and delivery of the irrigation equipment. Most energy for irrigation is used for groundwater pumping. As groundwater irrigation, in general, provides greater flexibility than other types in responding to fluctuating water demands, its relative importance is likely to increase.

> Irrigation is accounting for 40% of all food production.

The electric energy required for pumping is dramatically illustrated by the situation in India (see also section 5.3). India has around 21 million pumps connected to a primarily fossil fuel grid and almost 9 million pumps are diesel operated. These pumps contribute 8%–12% of the total annual greenhouse gas emissions in India. Subsidized energy has accelerated a perpetual cycle of increasing water and energy use. In some parts of the country, the water tables are receding at an alarming rate, thus requiring even more energy for pumping purposes.

Solar powered irrigation systems (SPIS) can provide reliable, cost effective, and environmentally sustainable energy for decentralized irrigation services in a growing number of settings and are discussed in more detail elsewhere.[279] While solar-based pumping is becoming competitive compared to traditional pumping, the financing structure must be different, since the up-front investment is higher, while 'fuel' is free. Life cycle assessments of emissions show a reduction potential of $CO_{2\text{-eq}}$/kWh as much as 95–98% of SPIS, compared to grid connected pumps (where electric power comes from fossil fuel power sources) or diesel pumps.[280]

It is recognized that solar-based pumping causes new risks for long-term water resources. Since the operational cost of solar PV pumps is negligible, it could result in overdrawing of water. To combat that unintended consequence, government programmes generally support pumps with low capacity, or farmers who accept subsidies to purchase solar water pumps must switch to drip irrigation. Since solar panels generate intermittent electricity some water or battery storage may be required.

The agricultural sector needs accurate, reliable, and timely weather and climate information for daily decisions and long-term planning. The Global Framework for Climate Services (https://gfcs.wmo.int/) has been formed, supported by WMO to supply this kind of help

7.5.2 Fertilizers

N (nitrogen), P (phosphorus), and K (potassium) are the three primary nutrients of plants. N is an important component of proteins, and as such is an essential nutrient for plants. P is a key to energy transfers in plants and is also a component of nucleic acids and lipids. K has an important role in plant metabolism, such as photosynthesis, activation of enzymes, and osmoregulation (the regulation of the osmotic pressure of an organism's fluids; to keep the organism's fluids from becoming too diluted or too concentrated). Unlike P and K, N does not persist in the soil long after application, making frequent reapplication of N-based fertilizers necessary. NPK are not the only necessary nutrients. Calcium (Ca), sulfur (S), and magnesium (Mg) are other so-called macronutrients. Oxygen should not be forgotten. If the soil does not have sufficient air exchange it will turn anaerobic (see also section 7.6). Other trace minerals needed include iron (Fe), boron (B), chlorine (Cl), manganese (Mn), zinc (Zn), copper (Cu), molybdenum (Mo), and nickel (Ni). All the nutrients are naturally in ion form.

Fertilizers have had a dramatic impact on crop productivity, in particular after the World War II, and are an important factor in agriculture today. They are responsible for substantial increases in crop yields and allow crops to be planted in soils that would otherwise be nutrient deficient. The relative significance of fertilizers is increasing as the population grows and as more low-income countries increase their fertilization rates. However, yield increases from fertilized crops come at a cost; fertilizers are large energy consumers.

Nitrogen is essential for crop production, but for growth the plants cannot use N in the atmosphere since it exists in its unreactive form N_2. There are exceptions: legumes – such as peas, beans and other pulses – have the unique ability of being able to convert atmospheric N_2 into reactive nitrogen, which can then be used for crop growth. For most crops, atmospheric N_2 is converted to ammonia (NH_3), which is the reactive N source that plants can use. This is realized using the famous Haber-Bosch process to produce N fertilizers.[281] The growth in agriculture output, using fertilizers, genetic breeding, irrigation, and machinery has had a cost in terms of ecology, energy, and other resources.

Any organisms, including microorganisms, in biological nutrient removal or crops on a farm, need nutrients to grow. Deficient nutrients can be added

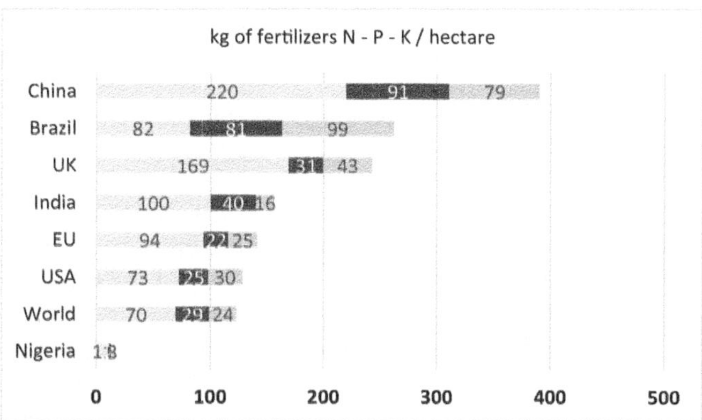

Figure 7.7 Fertilizer (left to right: nitrogen, phosphorus, potassium) use for cropland in some countries.[282]

either as synthetic fertilizers or as organic additions like manure. Figure 7.7 shows that there are huge differences between different parts of the world how much fertilizers are used. The difference between China and most sub-Saharan countries is striking.

Obviously, some countries need more nutrients to grow the crops, while others are over-applying the fertilizers. The nitrogen use efficiency (NUE) is defined as the ratio:

NUE = (nitrogen in the crop) / (nitrogen in the fertilizer or manure)

Ideally NUE would be 100%. The global average of NUE has been only 40–50% since 1980. This means than more than half of the N is not used by the crop but becomes a pollutant. There are great differences between nations. China and India have NUEs of 30–35% while Western Europe and the USA have of the order 50–70%. However, to make a fair comparison between countries, many factors have to be considered, like differences in climate, environmental conditions, vegetation, and soil types. Some countries like Democratic Republic of Congo, Tanzania, and Nigeria have NUEs exceeding 100%, indicating that they are under-supplying N. The crop does not get sufficient nutrient from the fertilizer, so it consumes N from the soil. This is called nitrogen mining. This will deplete the soil and deteriorate the food production in the long run.

> The global nitrogen use efficiency is <50%

There are plentiful suggestions about how to create economic incentives for better fertilizer practice. Fertilizers might be more expensive in countries that overuse them. In countries with low yields and small fertilizer use, however, fertilizers might be subsidized.

Fertilizer production consumes approximately 1.2% of the world's total energy on an annual basis. Ammonia production accounts for the bulk, or approximately 90% of the industry's total. Ammonia production has become much more efficient during the last 30 years, reducing the energy need – and carbon footprint – by some 30%.[283]

The Haber-Bosch process converts N_2 to NH_3 by adding hydrogen (H_2):

$$N_2 + 3H_2 \Leftrightarrow 2NH_3$$

In a typical ammonia production plant, natural gas provides the feedstock for hydrogen gas production via methane reforming and serves as a fuel for the plant. Natural gas consumption accounts for 50–75% of the total energy use of an ammonia production plant.

The global average of the energy consumption to produce 1 kg of NH_3 is:

- Global average: 41 MJ (=11.4 kWh)
- In industrial countries: typical range 33–36 MJ (=9.2–10 kWh)
- In low-income countries: 36–46 MJ (= 10–12.8 kWh)
- Using the best available technology: 23.5 MJ (=6.5 kWh).[284]

> Energy for 1 kg
> of N-fertilizers
> (kWh):
>
> Production ~11
> Transport ~0.3
> Spreading ~0.8

Energy for the transport and spreading of the fertilizers is usually of the order one tenth of the energy needed to produce the fertilizers. The energy required to produce phosphorus fertilizer is around 10 kWh/kg P.

7.6 CONSERVATION AGRICULTURE

In section 5.8 we asked the question if 'nature can trust us'. Our food habits had been questioned 50 years ago by Frances Moore Lappé (see section 7.1). In section 7.2 we reminded that livestock farming, and cattle grazing can use other types of land than arable farming. Any farmer knows this. Regenerative farming is based on various agricultural and ecological practices, with a particular emphasis on minimal soil disturbance and applicable local practices.

The US farmer Joel Salatin became a celebrity, portrayed in the book *The Omnivore's Dilemma*, authored by Michael Pollan[285] that received a lot of attention. Pollan's simple question was 'What should we have for dinner?'. The book challenged the way we view the ecology of eating. He traced food chains that sustain us, from the source to the table. This became an alarming reminder of how our food choices impact the health of our species and our planet. What's at stake in our food preferences is not only our own health, but the health of the environment

> What we eat
> affects our
> wellbeing and
> the health of the
> environment

that sustains life on earth. Pollan's question has a profound consequence for all of us, politically, economically, psychologically, and ethically.

Around a decade later, David R. Montgomery travelled around world to study how farmers have been able to bring ailing soil back to life remarkably fast. In his 2017 book *Growing a Revolution*[286] he described farming in both the high-income and low-income world. A new combination of farming practices can deliver innovative, cost-effective solutions to problems farmers face today.

There have been many debates about conventional and organic or regenerative farming. Montgomery describes how and why regenerative or conservation agriculture help restore soil health and fertility. Innovative farmers have shown that it is both possible and profitable to stop ploughing up the soil and blanketing fields with chemicals. Soil biology has often been overlooked while chemistry and physics have been emphasized. Farmers stress the importance of 100% groundcover all year around. In a climate with distinct rainy and dry seasons, the grass has to decay biologically during the dry season, before the next growth season. It is crucial to have as much green plant cover as possible all year long.

The benefits of groundcover include moisture retention, rainfall infiltration, weed prevention, increased organic matter, and soil erosion control. Improving soil structure will increase the ability of the soil both to infiltrate and to store water. Two parameters are crucial for the water holding capacity:

> Water holding capacity depends on soil particles and on carbon content.

- Soil particle sizes, and
- Carbon content.

Soil particles have a capillary film around them that will hold water, so the more particles, the greater the water-holding capacity. The photosynthetic process is the key process that transfers atmospheric CO_2 into the soil. The carbon content in the soil is a measure of its fertility and increases the water-holding capacity. Farmers do not talk about *rainfall* but about *effective rainfall*. This is a measure of how much rainfall has infiltrated the soil and then stored via organic matter.

As carefully described by Elaine Ingham,[287] a healthy soil contains a balance of microorganisms, protozoa, fungi, pathogens, and predators. The photosynthetic process to regenerate soils is a crucial tool for carbon sequestration, a biological tool as opposed to technological carbon capture and storage, described in section 6.10. A healthy soil should have an aerobic condition, an oxygen level of at least 6 mg/litre. This is achieved if the soil allows air to flow in, thus favouring aerobic microorganisms. Fungi and bacteria have considerably more nitrogen in their bodies than other organisms. Along with nitrogen they contain other critical plant nutrients—high levels of phosphorus, potassium, sulfur, magnesium, calcium, and so on. In contrast, an unhealthy soil lacks air exchange and turns anaerobic. This happens, for example, when compaction takes place. Water cannot penetrate. It loses its nutrients. Many disease-causing organisms are anaerobic. When chemical fertilizers are applied, salt will penetrate the soil and kill important parts of the microbial life.

> A healthy soil is aerobic

The conclusion from work on regenerative farming is that the combination of no-till planting, cover crops, and diverse crop rotations provides the essential recipe to rebuild soil organic matter. Farmers using these practices cultivate beneficial soil life, smother weeds, and suppress pests while relying on far less, if any, fertilizer and pesticides. Soil research studies have concluded that increasing cultivation leads to a loss of carbon-rich macroaggregates (collections of particles made up from organic matter and clay) and an increase of carbon-depleted microaggregates in soils.[288] Macroaggregates provide a key source of organic matter for roots as well as for bacteria and fungi. Some sources call the regenerative agriculture 'unconventional'. However, it is argued that the regenerative operation was the conventional until a century ago and got replaced by the 'unconventional' use of chemical fertilizers.

Practicing regenerative agriculture, Montgomery claims, is good for farmers *and* the environment. Regenerative practices allow the farms to use less water, use less fossil fuel and agrochemicals, generate less pollution, lower carbon emissions – and stash carbon underground. Still, they manage to maintain crop yields. Regenerative farming looks like a combination of ancient wisdom and modern science.

So, what happens? In conventional farming the soil is ploughed and emits CO_2 to the atmosphere and becomes more vulnerable for erosion. Furthermore, microorganisms making the soil fertile will die and artificial fertilizers must be used. In regenerative farming, ploughing is minimized. The result is that more carbon is stored in the soil.

> Ploughing causes CO_2 emission from the soil

Another feature of regenerative farming is that cattle grazing is integrated with arable farming. By having cattle on grass pastures, the cattle can trample organic material in the soil and provide organic fertilizers. It has been claimed that cattle will cause desertification. However, the opposite has been demonstrated. Allan Savory[289] argues that the single most important action is to have herds of grazing animals that can improve the soil and prevent desertification. The cattle should be moved between different pastures. Like Savory, Montgomery claims that meat from a well-maintained regenerative farm can be more climate friendly than vegetarian food from an industrial conventionally operated farm. The methane emission from cattle can be compensated by the carbon added to the soil. However, this conclusion from Montgomery is also a topic of a lot of scientific debate and several parameters must be considered. One important factor is what the animals are fed with.

> Cattle grazing and crop growing are integrated

Further considering the land use discussion in section 7.2: globally there is twice as much natural pasture as arable land. Crops cannot be produced in the natural pastures, but ruminant animals can eat the

> A cow can eat grass that people cannot eat. The benefit to have a cow is lost if it is fed with cereals and soya.

grass growing there. Therefore, cattle, sheep, and goats have a natural role to play in the food provision in the world. The competitive edge of these animals is that they can eat grass that people cannot eat. The reason to have a cow is lost if it is fed with cereals and soya. Pigs and chicken are much better at converting these feeds to food.[290]

Regenerative farming alone does not solve the climate crisis.[291] This must be done primarily by not burning more fossil fuel. However, turning the world's grasslands into healthy grassland is an important tool to handle the climate and food crisis. The carbon addition to the soil is a crucial contribution to a more sustainable agriculture and more food to the world.

7.7 FOOD-RELATED HEALTH

Water, including good water quality, is crucial for food production. Many food-borne illnesses can be traced back to poor-quality water in food production, processing, and preparation. Food production also generates wastewater, and too often effluents are disposed improperly into land and water ecosystems. The produced (used) water may contain not only nutrients and organic compounds but also pathogens. This affects rivers and lakes as well as biodiversity.

Food supply chains have been identified as one of the largest global environmental threats, and unhealthy diets are a most important risk factor for several diseases.[292] Food production is being compromised by extreme weather. Many countries are already experiencing falling yields. Negative impacts of climate change tend to outweigh potential positive impacts on national nutrition and food security through varietal breeding, improved farming practices, and reductions in poverty.

7.7.1 Targets for food production and consumption

The EAT-*Lancet* Commission has defined several universal scientific targets for the food system that apply to all people and the planet,[293] summarizing experiences from 37 leading scientists representing human health, agriculture, political sciences, and environmental sustainability. The global food system must operate within boundaries both for human health (diets) and for the final consumption and sustainable food production.

Food scientists have defined a safe operating space for food systems. This looks like a doughnut, so is known as doughnut economics (discussed in more detail in section 8.3 and illustrated in Figure 8.1). The wedges represent either dietary patterns or food production. Reaching outside the doughnut means, for example, high risks of biodiversity loss, insufficient vegetable intake, or increasing the risk of harm of the environment and/or human health. Operating inside the inner wedges – in the doughnut hole – will push humans into an uncertain and dangerous zone of hunger and malnutrition. For example, a diet rich in plant-based foods and with fewer animal source foods confers both improved health and environmental benefits. Overall, the literature indicates

> Half the healthy food plate should consist of vegetables and fruit and fruit

that such diets are 'win-win' in that they are good for both people and planet (compare section 5.6).

The Commission recommends that a plate, healthy for both humans and planet, should consist of around half a plate of vegetables and fruits; the other half, displayed by contribution to calories, should consist of primarily whole grains, plant protein sources, unsaturated plant oils, and (optionally) modest amounts of fish, meat, and dairy foods. Naturally, eating patterns are different but the recommendations can be applied in most cases. To follow these recommendations, the consumption of fruits, vegetables, and legumes must almost double, while consumption of sugars and red meat (mostly in rich countries) should be reduced by half. The impacts on the proposed diet would have huge impacts on both human and planet health.

7.7.2 Health risks factors
7.7.2.1 *Human health*
Certain groups of people in low-income countries are especially at risk for adverse health effects from climate change. These at-risk groups include urban people living in poverty, older adults, young children, traditional societies, subsistence farmers, and coastal populations.

- Climate change impact on agriculture and other food systems can increase rates of malnutrition and food-borne illnesses.
- Infectious diseases can increase. The spread of meningococcal (epidemic) meningitis is often linked to climate changes, especially drought.
- Mosquito-borne diseases such as malaria, dengue, and West Nile virus may increase in areas projected to receive more precipitation and flooding.

7.7.2.2 *Antibiotics*
Antibiotics are used regularly in animal feed at a rate of 2–50 g/ton. A substantial portion of this is used to promote growth and not for the treatment of infections.[294] It is assumed that microorganisms present in the animal feed consume a large portion of the nutrients in the feed. They also inhibit absorption from the intestine and produce toxins that can harm the health of the animals. The improved growth due to antibiotics might stem from their ability to suppress these harmful organisms.

Up to 18 000 tons of antibiotics is used annually for the livestock or 2000 kg of antibiotics every hour. The potential of resistant bacteria is apparent. These bacteria proliferate in the animal. Through interaction, the resistant bacteria are transmitted to other animals as well as to human beings, thus leading to colonization by antibiotic resistant bacteria. The faeces of the animal, as well as from humans, often contain the resistant bacteria. Transfer of bacteria from animal to human is possible through many practices that occur on farms and in slaughterhouses. Since antibiotics are used to favour the animal growth and compensate for insufficient hygienic conditions, the economic interests in the meat industry are huge.

> Too often antibiotics are used for growth rather than for infections

Then somebody else must pay for the costs of deteriorating food quality and resistant bacteria.

7.8 ACTIONS NEEDED

As expressed clearly by the EAT-*Lancet* Commission, a radical transformation of the global food system is urgently needed. Food production must become sustainable, both to improve our health and to save the Earth. Some people eat too much, others are hungry. Food has a huge impact on the climate and the opposite: climate change has a grave impact on food production.

How to reduce the carbon footprint from agriculture and food production is not obvious. Fertilizers are still needed and cattle that produce methane will remain, but new ways of farming are being tried out (section 7.6). Energy production has more apparent alternatives, like renewable energy. We will need many types of solutions like dietary changes, reduced food waste, higher agricultural efficiency, and methods that can produce scalable and affordable food alternatives. All of this requires systems thinking.

Food production must not only produce enough food to feed the global population. Agriculture must also produce healthy food where the production does not destroy the planet. The world must get away from the large-scale deforestation to produce food and feed, also resulting in loss of biodiversity. As discussed in section 7.6, when a rainforest is cleared, burned, and the land subjected to annual tillage and burning, it has often been observed that this once highly productive landscape now barely able to support a maize crop. What happened? There is a growing understanding that the answers are found in the abundance and diversity of life hidden below the surface. There are many attempts to restore agriculture productivity in degraded lands and adopt sustainable production practices.

Too many crops today are used for animal feed. Still, it is important to consider local conditions for livestock production. These are some of the factors to consider:

- Dietary shift will have a great impact on both human health and ecology.
- Producing biofuel competes with food production. This coupling must be recognized. Land use must be carefully planned to make the best use of productive soil.
- Too much food is wasted along its path from production to the table.
- Water supply is crucial for food production. This requires wise management
- Water is polluted from agriculture. The role of fertilizers must be recognized.
- A lot of energy is used for food production. New farming methods must be considered and practiced.

doi: 10.2166/9781789062908_0159

Chapter 8
Economics

In 2021 the world's richest 1% have more than double the wealth compared to 6.9 billion people or 88% of the global population. In countries around the world, small elites are taking an ever-increasing share of their nation's income, while hundreds of millions of people are still living without access to clean water and without enough food to feed their families.

Producers and workers who grow, process, and pack our food are getting a smaller and smaller share of the consumer price. In some cases, this means that those who produce our food are themselves going hungry.

Tea. Rather tea dust. A woman shows the tea that the tea workers get as a monthly ration. It simply passes through the tea strainer, she tells us. Though these are the hands that have built the plantation and toil to bring us the tea that we relish, they hardly get any share of it.

Permission from Oxfam International (www.oxfam.org) gratefully acknowledged. (Photo: Roanna Rahman/Oxfam)

For a long time, we have realized that unlimited growth is impossible in a finite world. This was made obvious by the *Limits to Growth* debate in the 1970s, as mentioned in the prologue. Still many people, including economists, think of growth as a linear phenomenon. Asking anybody around me how many years it would take to double capital from 100 to 200 with 2% annual interest rate, many persons guess around 50 years. Any engineer will recognize that we talk about exponential growth, so the doubling time is only around 35 years. That wealth is a source of well-being is one of the core principles in our consumer-oriented societies and one of the basic axioms of development policies.

In wealthy countries we have never been more affluent. There is something quite extraordinary that so many people and decision makers believe that continued economic growth is the solution to our problems. Are we willing to sacrifice any of our well-being to reach the climate goals?

The climate crisis cannot be handled by singular actions like paper recycling, electric cars, or replacing plastic straws with paper straws. If we do not 'connect the dots' there is no chance of reaching the goal to limit global warming.

The IPCC report, published in 2021 before the COP26 conference naturally got a lot of media attention. A leading politician in my country probably represents a typical attitude of having a blind spot. The essence of the statement was that 'the climate crisis demands that we take radical measures to mitigate the global warming. .... Furthermore, we are really satisfied that charter flights now – with less restrictions due to the pandemic – can operate again to the Mediterranean countries and to Florida.'

The global responsibility
A fundamental outcome of the Paris agreements in 2015 is the *principle of equity*[295]. It means that wealthy countries – that have caused most of the emissions – should lead the huge challenge to limit the global warming by reducing national emissions and adapt to the impacts of climate change. This is fundamentally important for poor countries to be able to develop the basic infrastructure that we simply take for granted in wealthier countries, such as hospitals, health care, clean water, and affordable energy for everybody. Furthermore, too many people lack the access to information and technology.

> Wealthy countries should lead the challenge to limit the global warming

Countries that ratified the Paris Agreement in 2015 had to submit a Nationally Determined Contribution. These are published on the UNFCCC website[296]. Most of the Nationally Determined Contributions contain a plan to limit the GHG emissions and a description of the funding to carry out the plans.

8.1 INEQUALITIES

The dramatic increase of fossil fuel use has caused the global warming crisis. In Chapter 4 we listed some of the dramatic consequences of climate change. Countries will lose their islands because of sea level rise and extreme events. The destruction of biodiversity is threatening nature, cultures, and traditions. Buildings are damaged and jobs are lost due to extreme weather. It has been argued that rich countries should pay for the climate-induced losses and damages that poor countries are suffering. The results from the COP26 conference did not include any decision about climate finance to address the rising costs of losses and damages in low-income countries. Rich nations said they (we!) would establish 'a dialogue' to discuss 'arrangements for the funding of activities to avert, minimise and address loss and damage'.

At the COP26 meeting, Prime Minister Mia Mottley from Barbados expressed the frustration from small island states: asking countries on the frontlines of the climate crisis to pay for climate damages is 'like asking the passengers of a car crash to pay for damages, rather than the driver'.

Poverty and inequality are not just a result of bad luck. They are most often a result of political decisions and underlying structural factors. Professor Joseph E. Stiglitz was Chief Economist at the World Bank until early 2000 and won the Nobel Prize for economics in 2001. He demonstrates that macroeconomic policies and tax systems serve the interests of the top, often at the expense of the rest of society[297]. These inequalities, between nations and between people, will amplify the negative impact of conflict, climate impact and extremes, and economic downturns. Globally, poverty has declined but income inequality has remained high during the last two decades and is actually rising in nearly half the countries in the world[298]. As Stieglitz points out: societies with more economic inequality tend to have more political inequality.

> Income inequality has remained high during the last two decades

Inequality includes more aspects than distribution of income or wealth. It also relates to access to nutritious food and to social and health infrastructures. It implies that economic growth alone will not eliminate poverty, food insecurity, and malnutrition. Actually, income inequality can also result in overweight and obesity. The poor may not afford nutritious food and tend to eat cheap, energy-dense, and nutrient-poor food.

The world is facing a serious risk from the combination of growing social inequality and increasing ecological degradation. This is a powerful recipe for collapsing civilization as described by the environmental historian Jared Diamond[299]. He argues that if the short-term interests of a decision-making elite do not embrace the long-term interests of the whole society, there is a 'blueprint for trouble'. This risk had been pointed out much earlier by the landmark books *Limits to Growth*[300] and *Small is Beautiful*[301].

> The world faces a growing social inequality and increasing ecological degradation, a 'blueprint for trouble'.

As of March 2021, there were 2755 US$ billionaires in the world, an increase of 660 in one year. They saw their fortunes grow more during Covid-19 than they have in the whole of the previous 14 years. Their net worth value was US$ 13 100 billion (up from US$ 8000 billion during one pandemic year), or an average wealth of US$ 4.75 billion[302]. Just under half of the billionaires are from the USA and China. Imagine a wealth tax of 1%. This would generate US$130 billion that could be used to address some of the global issues like education for all children, health service for everybody, clean water, and clean energy. Compare this amount of money with the annual climate fund of US$100 billion, promised by wealthy countries to be realized in 2020, but now postponed (section 3.6). Another contributor to global inequality is the global corporate tax system allowing multinational corporations to use tax loopholes and tax havens. Changes may become inevitable.

> 1% of the wealth of the world's billionaires would bring US$130 billion.

There is no shortage of money in the world. As a response to the pandemic, the governments in the world released US$16 000 billion[303].

One example, presented at the Climate Vulnerability Forum[304], may illustrate the cost of climate mitigation. To protect a shoreline with water breakers and embankments is extremely expensive. To protect 1 metre of shoreline in the Maldives with breakwaters and embankment would cost US$7000. And the Maldives contains hundreds of islands.

In previous chapters we have highlighted some of the absurd inequalities in the world, between nations and between people. Here we will repeat some observations.

8.1.1 Climate change

It is recognized that the rich nations have caused most global warming, while mostly poor nations suffer from the impact of climate change. Our media usually pay more attention to disasters in wealthy countries than in poor countries. The contrast between Figures 4.2 and 4.3, and Figure 4.4 (repeated here) illustrates the differences in vulnerability.

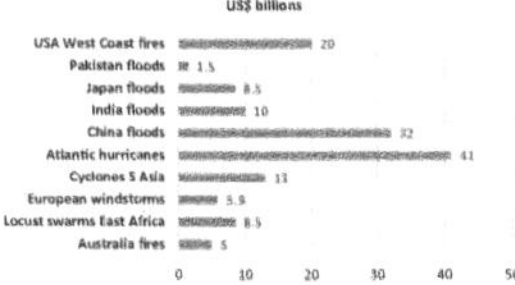

Figure 4.2. Disaster costs.

Figure 4.3. Disaster costs/GDP

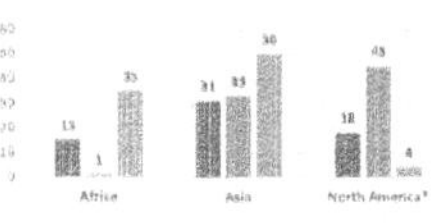

Figure 4.4. Weather related disasters.

8.1.2 Water inequalities

Water availability is truly different around the Earth. Rich nations can compensate for water scarcity while people in poor nations suffer, to some extent illustrated in Figure 5.5. The glacier losses will hit hard, Figure 5.6, and numerous vulnerable people living on the slopes below glaciers will suffer from increasingly insecure water supply.

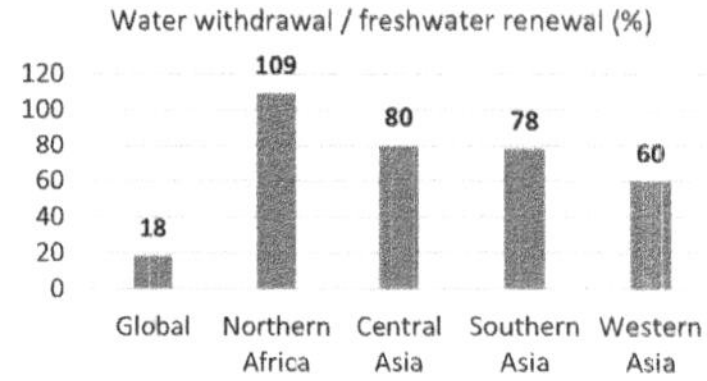

Figure 5.5. Water scarcity

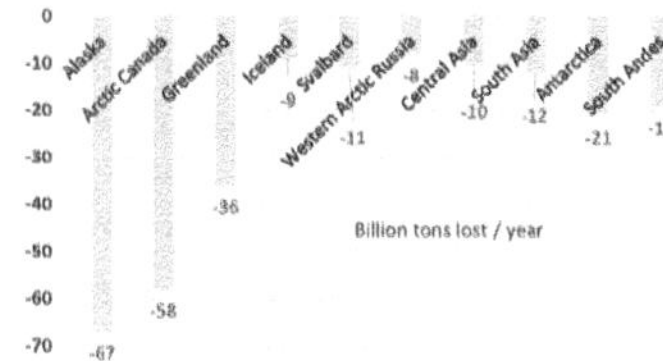

Figure 5.6. Glacier losses

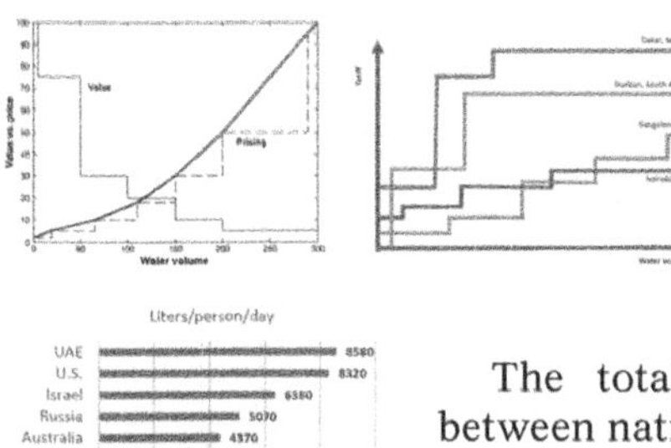

How water is valued often determines the consumption of water. Even the poorest should be able to afford basic water needs (Figures 5.8 and 5.9) while the extravagant water user should pay properly. Progressive tariffs are an important tool to achieve this.

The total water footprint is extremely unequal between nations (Figure 5.12) and between people.

8.1.3 Energy inequalities

The historical records are obvious (Figure 3.1). The accumulated emissions are clear indicators of how much fossil fuel has been used in various countries. Fossil fuel use has been the driver to build up the wealth of nations. Now, we in the wealthy nations seem to wish telling poor nations that they cannot use the same tools to build up their wealth.

The inequalities between nations of both primary energy (Figure 6.10, left) and electrical energy (Figure 6.11, right) consumption are bizarre. Wealthy nations are telling nations having an energy consumption of 1–2% compared to the rich ones: 'you have to mitigate global warming'.

Air pollution caused by fossil fuel burning is a frightening killer (Figure 6.22). More of the premature deaths happen in poor countries, where coal is a major source of power, both indoor and outdoors.

Critical minerals and metals for energy storage, such as cobalt and lithium, are found mostly in not-so-wealthy countries (Figure 6.21), while their dominating use is in high-income countries. Currently there are heated debates in Europe if mines for critical metals should be opened in the countries where the metals are meant to be used.

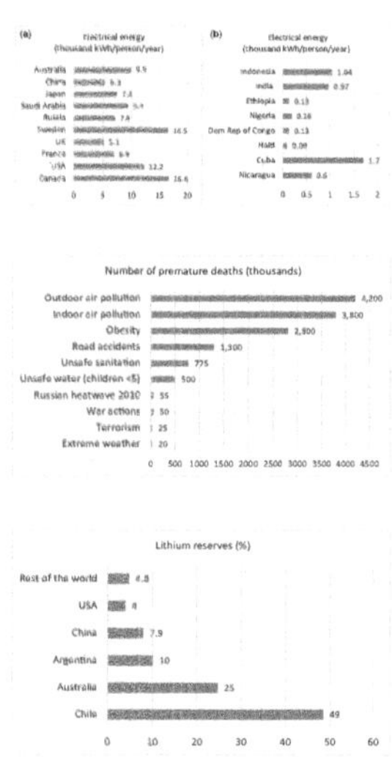

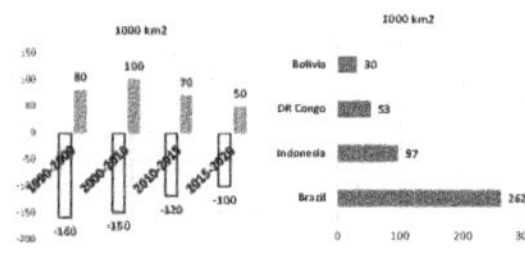

8.1.4 Land use and deforestation

The huge development of agricultural land, mostly aimed for animal production, has caused serious deprivation of land. Deforestation is a serious threat to biodiversity, and ultimately our own health. A few countries are hit hard (Figure 3.3), while other nations have enjoyed the (short-term) fruits of this development.

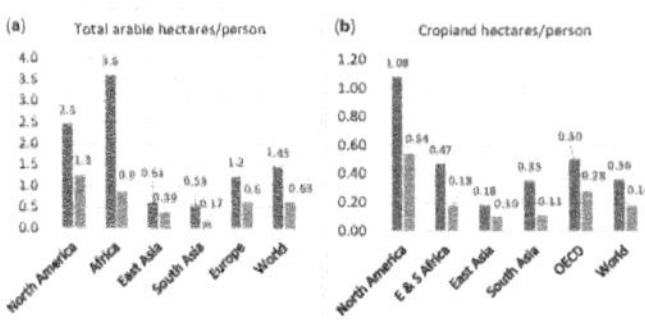

The decrease of total arable land (Figure 7.5a) and total cropland (Figure 7.5b) per person is showing huge differences between the continents concerning available land per person.

8.1.5 Food inequalities

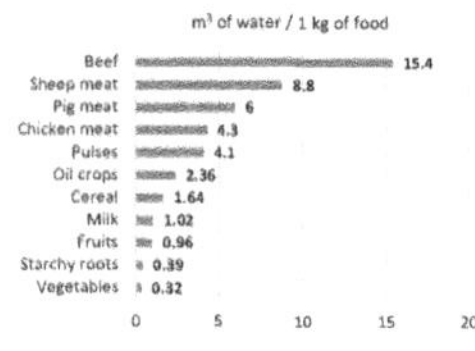

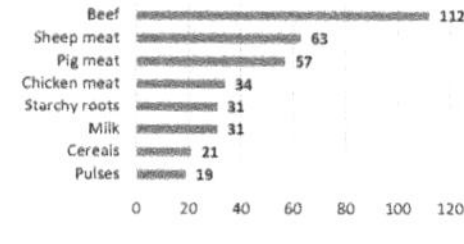

Generally speaking, rich people eat more meat products. This is an expensive way to get calories and nutrients, and the water requirement is large (Figures 5.10 and 5.11).

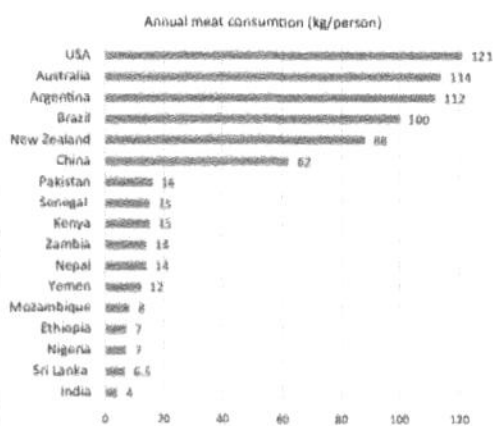

The differences in meat consumption between nations is striking (Figure 7.3).

8.2 ECONOMIC CONSEQUENCES OF CLIMATE CHANGE

The economic consequences of climate change have been studied extensively during the last couple of decades[305]. Several economists have warned that our responses to global warming have been inadequate[306] and others have demonstrated that current models underestimate the cost of carbon emissions[307]. Timing is crucial. The process of endlessly debating climate policies comes with enormous costs. One of the most engaged economists on climate and inequality issues has been Joseph Stiglitz, also a lead author of the 1995 IPCC report.

> The process of endlessly debating climate policies comes with enormous costs.

A major challenge that many economic models do not take into consideration is the feedback loop between economy and climate:

- Greenhouse gas (GHG) emissions are a by-product of economic activity.
- The amount of GHG emissions per capita depend on a country's level of economic development, as well as government policies to reduce emissions.
- GHG causes global warming and climate change.
- Climate change has profound economic impacts.

8.2.1 The polluter pays principle

The 'polluter pays' principle (PPP) has been accepted and practised for 30 years. It simply states that those who produce pollution should bear the costs of managing it to prevent damage to human health or to the environment. It was formulated in the 1992 Rio Declaration[308] as Principle 16:

'National authorities should endeavour to promote the internalization of environmental costs and the use of economic instruments, taking into account the approach that the polluter should, in principle, bear the cost of pollution, with due regard to the public interest and without distorting international trade and investment.'

So, why is the PPP not applied for discharge of CO_2 or other greenhouse gases? Since societies, particularly in wealthy countries, have been slow to recognize the couplings between human activities and global warming, emitters are generally not held responsible for handling this form of pollution. Since the costs will not be carried by the emitters they are **'externalized'** to society. The atmosphere is described a 'global commons' as everyone shares it and has the right to use it. Until people realize that discharging CO_2 in the air is like dumping toilet waste directly into a river, they do not want to pay what it costs.

> Until people have the idea that releasing CO_2 in the air is like releasing toilet waste directly into a river, they do not want to pay what is costs.

Rich countries are responsible for almost 80% of historical emissions. Low-income countries – where most of the victims are – argue that rich countries should pay for the externalities. Wealthy nations are concerned that they could be held liable for the pollution they have caused. When loss and damage was included in the Paris Agreement in 2015, the USA pushed for a clause to be added which stated that the accord 'does not involve or provide a basis for any liability or compensation'.

8.2.2 Paying for climate change

The understanding of the damage caused by climate change has increased a lot recently. Today there is plenty of real-life data about issues, such as the relationship between temperature and human wellbeing. The unequal impacts of climate change in different locations can also be considered. The cost for greenhouse gas discharge is becoming apparent and painful, for example:

- Small island countries lose their islands because of sea level rise and extreme events.
- Indigenous people lose their culture, link to the environment, and identity because of deforestation, mining operations, or oil exploration.
- Lives are lost and buildings are damaged from extreme events.
- Droughts are devastating large regions.
- Glacier loss risks the water supply for millions of people.

This raises the question as to what the real cost of climate change is. This important number is debated. It is argued that it should include not only costs for global damage but also the social cost of carbon (SCC) and the impact of today's emissions on future generations. The landmark report *The Stern Review* from 2006[309] concluded that 'the benefits of strong and early action on climate change far outweigh the costs of inaction'. The report of more than 700 pages is probably one of the most influential reports on climate change ever produced.

> The cost of climate change should include costs for global damage, the social cost of carbon, and the impact on future generations.

The Stern Review emphasizes that climate change has some features that distinguish the climate change externality from other externalities:

- It is *global* in its causes and consequences.
- Impacts of climate change are *long-term* and *persistent*.
- *Uncertainties* about the economic impacts are pervasive.
- There is a serious risk of major, *irreversible change*.

Nicholas Stern and Joseph Stiglitz (see section 8.1) have collaborated in a report[310] claiming that the way that the social costs of carbon have been calculated by economists, using a framework called Integrated Assessment models (IAMs) is not adequate. The IAMs overestimate the costs of climate action and underestimate its benefits. Thus, the SCC is presently valued in the U.S. at US\$51 per ton of CO_2. Stern and Stiglitz argue that the cost should approach US\$100 per ton by 2030. Whatever the price will be: if the price goes up, emissions will hopefully go down.

> How expensive is US\$100/ton or US\$0.1/kg of CO_2? The CO_2 emissions from 1 liter of fuel are:
>
> - gasoline: 2.351 kg,
> - diesel: 2.693 kg.
>
> The carbon cost would increase the price US\$0.24 – US\$0.27 per liter.

The European Union and a dozen US states operate carbon markets. Almost half of the biggest 500 companies in the world have an internal carbon price they use to test the viability of new projects. Almost 30 nations have carbon taxes, which range from less than US\$1/ton in Mexico to about US\$140/ton in Sweden. Today only about 4% of

> About 4% of emissions have a CO_2 price >US\$40/ton

emissions are covered by a carbon price above US\$40/ton[311]. Many countries use permit trading alongside targeted taxes on dirty fuels such as coal. Still, carbon pricing covers only about 20% of global emissions. Twelve of the G20 countries have established nationwide prices on CO_2 emissions[312]. The highest taxes are established in Italy (US\$67), France (US\$60), and the UK (US\$58), while Australia, China, and the USA taxes are only US\$12, 6, and 6, respectively.

As a preparation for the COP26 meeting, New York University's Institute for Policy Integrity made a survey among 738 economists from around the world[313]. They found 'a strong level of consensus' among the economists of the economic importance of climate action. In a survey in 2015 they had found that about half of the economists strongly agreed that drastic actions should be taken immediately. In 2021 some 75% of the respondents had this view.

8.3 DOUGHNUT ECONOMICS

The boundaries of economics and its relation to humanity and to the global resources have been illustrated in a single image, a 'doughnut'. This brilliant picture illustrates the state of the earth and the humanity in a single image. Doughnut economics has been presented by Kate Raworth[314] and expresses how social and planetary boundaries should be the guide for all economic thinking to recognize that we live in a limited world. She emphasizes that mainstream economists mostly ignore ecological

> Social and planetary boundaries should be the guide for all economic thinking

stress, deforestation, water scarcity, and other global problems to the periphery of economic thought, until they become so severe that they are damaging the economy. The idea of economic systems where profit is the dominating economic performance index must be replaced.

The *inner* boundary of the doughnut defines the social foundation of wellbeing. The doughnut hole indicates critical human deprivation where people are falling short on life's essentials, such as food, water, healthcare and political freedom of expression. We should prioritize that everyone would be out of that hole. The *outer* boundary of the doughnut indicates the ecological ceiling. Overshooting the doughnut's outer crust means that we will risk a stable climate, healthy oceans, clean water, critical resources, and biodiversity on which all our wellbeing fundamentally depends. The space for humanity and life on earth is located between the inner and outer boundaries of the doughnut, Figure 8.1.

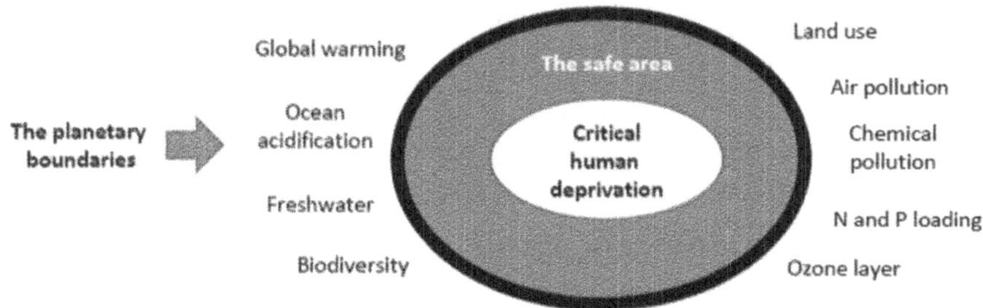

Figure 8.1 The doughnut picture of an economy for people (inspired by Raworth, 2017; Rockström *et al.*, 2009).

Already in 2009, a scientist group led by Johan Rockström[315] defined nine planetary boundaries. They define the outer edge of the doughnut, Figure 8.1. Getting outside the edge increases the risk of generating large-scale abrupt or irreversible environmental changes, leading to unacceptable environmental degradation and the risks for climate tipping points (see section 4.11) are apparent. If we limit our societies within the boundaries, humanity can continue to develop and thrive for generations to come. For some people this is intuitively obvious, for others it is a radical economic idea. The nine planetary boundaries are listed below (the numbers within the brackets show the chapters where they are discussed):

- Global warming and climate change (Chapters 4–5)
- Ocean acidification (6.14)
- Freshwater withdrawals (5.2, 5.3, 5.6, 6.4, 6.6, 6.8, 7.4)
- Biodiversity loss (2, 3.7, 5.8, 6.8, 6.13, 7.1, 7.6)
- Land conversion (3.7, 6.3, 6.5, 6.8, 7.1, 7.2, 7.5)
- Air pollution (4.7, 6.5, 6.14)
- Chemical pollution (6.3, 6.5, 6.6, 6.13)
- Nitrogen and phosphorus loading (5.3, 7.5)
- Ozone layer depletion (4.7, 4.10, 6.14).

The fact that each one of these planetary boundaries is discussed in several sections of this book indicates that the real problems are truly complex. The picture helps us to grasp the essentials, but we should not over-simplify the problems.

Doughnut economics emphasizes that we must get rid of the dominating measure of progress, the GDP. This measure does not give a true picture of the real economy, of resource utilization, or of the absurd inequalities in many societies and between nations. GDP is a key measure of short-term macroeconomic activity. GDP may capture the economic output that is generated by deforesting nature and producing cattle, but the degeneration of natural assets required to generate this productivity is not captured. National capital accounting is an attempt to include nature in the assessment of wealth. Assume that the production of meat has forced deforestation. Taking the national capital accounting of the rainforest as a basis for the price, it would be considerably more expensive to eat meat.

> Taking national capital accounting of rainforest as a basis for the price, it would be considerably more expensive to eat meat.

> GDP does not give a true picture of the real economy.

Already in the 1970s, Ernst Schumacher, as described in the prologue, had pointed out that we need another kind of economic thinking and sought to place ethics and human scale at the heart of economic thought. We have mixed up the need to grow with the need to thrive, or in Schumacher's wording: 'Any intelligent fool can make things BIGGER, more complex and more violent.

It takes a touch of genius – and a lot of courage – to move in the opposite direction'.

One of the tragedies of our consumerist society is the belief that a person's happiness depends on acquiring as many goods and much capital as possible. There is a long tradition of research around the relationship between wealth and happiness[316]. We often envision that wealth will lead to a stress-free life and therefore provide happiness. It is demonstrated that in *low-income* countries, there is a clear correlation between wealth and happiness but not as apparent in *wealthy* countries. Once people get rid of poverty, the correlation between income and happiness decreases. When we adapt to well-off living conditions the happiness that wealth brings us is no longer as great. Apparently, we need something more than growth.

8.3.1 Circular economics

Many decisions makers like to favour 'green growth'. Can we actually reach climate goals without sacrifices? A crucial part of the solution is found in circular economics. The throw-away economy is a crucial factor in the global warming. Like regenerative food production (section 7.6) any industrial production must be regenerative. Circular economy thinking involves a lot of ingredients[317]:

- Energy must depend on renewable sources, like solar, wind, waves, biomass, and geothermal sources.
- Waste should be eliminated by design.
- Food leftovers, mostly rich in nutrients, can be used to grow more food or feed.
- Used water is not wastewater, but a resource that can be recovered. It contains thermal and chemical energy and various valuable nutrients, like phosphorous.
- Toxic chemicals cannot be left in nature but have to be recovered or eliminated.
- Scrap metal becomes the source material for new production.
- As noted in section 6.12, critical metals and minerals must be recycled to avoid catastrophic extraction of them.

The purpose of the doughnut economics is to provide a new metric of the economic system where wealth is measured not only in GDP but human, social, ecological, cultural, and physical wealth.

8.4 ENERGY SUBSIDIES

Fossil-fuel subsidies are gigantic and are a huge barrier to shift from fossil to renewable energy sources. Already in 2009, the G20 countries pledged to gradually phase out fossil fuel subsidies[318]. In the years 2015–2019, G20 governments provided US$3.3 trillion ($10^{12}$) of direct support to coal, oil, and gas, and fossil fuel-fired power generation. The subsidies decreased from US$706 billion in 2015

> Fossil fuel subsidies are of the order US$1.7 billion/day or US$70 million/hour, around the clock

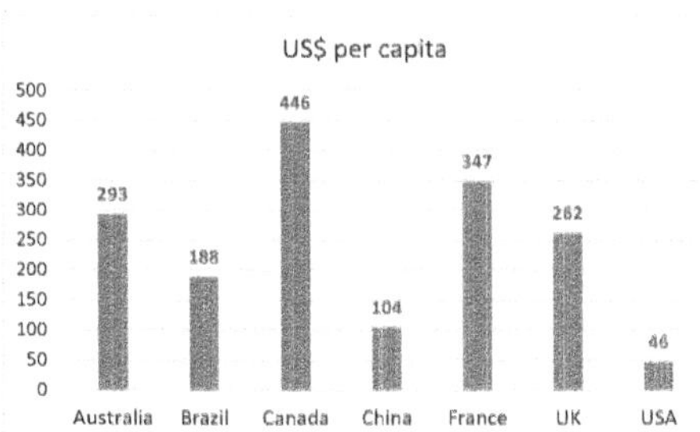

Figure 8.2 Fossil fuel subsidies per capita in some G20 countries in 2019[320].

to US$636 billion in 2019, only a 10% decrease. Governments, state-owned enterprises, and public financial institutions provide the huge subsidies for the production and consumption of fossil fuels. Sixty per cent of fossil-fuel support in 2019 went to producers and utilities and another large portion to consumer subsidies. This will disproportionately benefit wealthier consumers. Actually, coal subsidies have risen in recent years. IEA[319] has stressed that the world needs a 'historic surge in clean energy investment' to avoid severe impacts from climate change and accentuates that consumption subsidies for fossil fuels remain a great obstacle to reach a clean energy future.

The G20 countries produce 90% for the world's GNP and 80% of the greenhouse gas emissions. Subsidies for fossil fuels vary widely from country to country. The fossil fuel annual support per capita is notably high in some of the G20 countries as shown in Figure 8.2.

To get a feeling for the numbers: assume that the US$3.3 trillion, paid as subsidies for 5 years, would be used for solar energy investments. According to Wood Mackenzie's recently released US Solar PV System Price report, the average 100 MW utility scale system costs in 2020 were US$0.94/W. This would buy 3500 GW of solar capacity, an increase of the 2020 global solar capacity of 773 GW by a factor of 4.5.

Subsidies encourage consumption and tend to favour wealthier households that use more fuel and energy. It is an ineffective way to support low-income people compared to other ways of support. However, there are many examples of governments successfully initiating fossil fuel subsidy reforms protecting the vulnerable in effective social programmes, providing income support and health insurance schemes to poor households, and better funds for education[321]. Experiences from several countries demonstrate that subsidy and taxation reforms work better when they are part of a wider push for energy transformation. The pandemic in 2020 also motivated some countries to make some pricing reforms.

8.5 ACTIONS NEEDED

The climate crisis has made it apparent that the traditional capitalistic system, based on consumption and profit is not designed to cope with the actions needed to solve the climate crisis. Several changes must be made in economic thinking and in methods to limit the current emissions, such as:

- The growing economic inequality is a huge threat, not only to individual lives, but to the survival of societies and nations.
- Economics must take the social and environmental limits into consideration, the doughnut economics.
- Countries having caused most of the global warming should pay most of the cost to fight the climate crisis, an obvious application of the polluter pays principle.
- Emissions polluting the atmosphere must be given an adequate price tag.
- Subsidies to all fossil fuel production and consumption must terminate.

doi: 10.2166/9781789062908_0173

Chapter 9
Lifestyle

There can be only one permanent revolution – a moral one; the regeneration of the inner man. How is this revolution to take place? Nobody knows how it will take place in humanity, but every man feels it clearly in himself. And yet in our world everybody thinks of changing humanity, and nobody thinks of changing himself.

Leo Tolstoy

Never before have we had such an awareness of what we are doing to the planet, and never before have we had the power to do something about that… The future of humanity and indeed, all life on earth, now depends on us.

Sir David Attenborough

The IPCC reports are crystal clear: we need a transition and even a transformation. We need pervasive change of processes and behaviour at all levels: individuals, societies, industries, institutions, and governments. We simply must redefine our lifestyle and our consumption patterns and dare to think outside the box.

9.1 THE REAL DIFFICULTY

There is too much greed and too little compassion in the world. Science does not provide any solution how to decrease human greed and how to increase human compassion. This is the all-encompassing climate change challenge.

It is not global warming that makes the threat. It is the gasoline, the oil, the coal, the natural gas, the methane. This is caused by humans.

> The threat is not the global warming. The real threat is caused by our own choices of burning fossil fuels.

In the radio news on 6 January, 2015 it was reported that the oil price had dropped below US$50 per barrel. A striking comment was that 'the consumption of oil is increasing but not sufficiently fast, since there is a surplus production'. This is our economic imperative: grow or die. When the oil price had dropped there was hardly any discussion of the consequences for the climate. Most of us are aware of climate change, but this does not change our lifestyles or the political agenda, even if the whole earth is threatened. What is wrong with us?

As discussed in Chapter 8, the economic model built on consumption and growth is not designed to cope with the climate challenge. At the same time, we demand that politicians and decisions makers around the world should take a responsibility for systems issues that they cannot influence.

Climate change is becoming an existential threat and we should get prepared for it, not only in our private lives. Education of young people and academic curricula are hardly ever designed to prepare young people for the upcoming challenges but are designed along specialist silos. Civil engineers, urban planners, city managers, or architects are seldom trained to deal with climate change. Electrical engineers, together with other specialists, should be prepared for the necessary electrification of the society. Environmental engineers must discover all possible interactions that determine the fate of our nature. And so on.

Collectively, we have most of the technology to reduce greenhouse gases, but we have never really shown the will. Our value system, our lifestyle, mobility, what we eat, our clothes, our economic models – ***everything must work together*** to deal with the climate crisis. To seriously deal with climate change will probably require a completely new attitude to life. Many people have climate anxiety today – probably because we are not prepared to sacrifice anything? Maybe somebody else can start?

> *Everything must*
> *work together*
> to deal with the
> climate crisis.

The dominating groups choosing the train and eating less meat are found among young people, women, and people living in large cities. Climate change is causing an increasing worry, and more and more people make active choices to live a little smarter with respect to the climate.

9.2 RELATION BETWEEN WEALTH AND EMISSIONS

In section 8.1 we described some consequences of the huge economic inequalities in the world. The Oxfam Inequality report[322] presents not only the absurd inequalities in the world, but also clarifies with painful clarity how inequality kills. Gender and racial inequalities are increasing and widening economic inequalities are tearing the world apart. This is true not only between individuals but also between nations. The development is logical: structural policy choices are made for the richest and most powerful people.

The pandemic was a harsh reminder about the bizarre inequalities. With deprived access to vaccines, hospital care, public health, and food, millions of poor people died unnecessarily. At the same time, the richest people got even richer. Some Oxfam comments:

- The top 1% have achieved almost 20 times more wealth than the bottom 50% of the global population, since 1995.

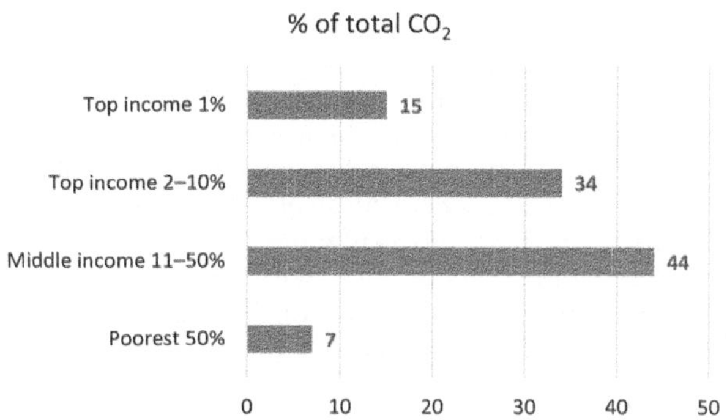

Figure 9.1 Contributions to global emissions from different income groups.

- 250 men have more wealth than all 1 billion women and girls in Africa, Latin America, and the Caribbean, combined.

The rich part of the global population is increasingly killing the planet. Figure 9.1 illustrates how the wealthiest people in the world are the primary contributors to global warming. The richest 1% emit 70 tons of CO_2 per capita and year, while the poorest 50% emit 1 ton. In 2019, CO_2 emissions per capita in the world was 4.93 tons per capita.

The top 1% wealthy people emit more than twice as much compared to almost 4 billion people at the bottom of the wealth pyramid. The high-income half of the global population produces 93% of all emissions, but it is usually the poorest part of the population that pays the cost for the consequences. The obvious outcome is, that unless the wealthiest 10% do not change their (our!) lifestyle the world will miss the 1.5 or 2°C goals, whatever the remaining 90% will do. Oxfam predicts that the contributions to the global emissions from the richest people will increase further during this decade.[323] The global crisis is caused by the richest people and only they can do anything about it, as discussed in section 8.1. The carbon footprint of the billionaires' houses, vehicles, aircrafts, and yachts are mindboggling, in excess of thousands of tons per year. Space tourism is the latest addition of exceptional carbon footprint from a few people. Their pollution to the atmosphere has no fee. The purpose is not to further explore space around us or to build on the technology for the good of humanity. It is luxury consumption and an exclusive adventure for a few. Our problem is that so many people admire the richest people.

Greta Thunberg, our Swedish climate activist has expressed it brilliantly at the Pre-COP26 Youth Summit: *'The climate crisis is of course only a symptom of a much larger crisis. A crisis based on the idea that some people are worth more than others, and therefore have the right to exploit and steal other people's land and resources. It is very naïve to believe that we can solve this crisis without confronting the roots of it'.*

Around half of the 1% richest people in the world come from the USA and China. This illustrates that emission reductions must take place not only domestically in rich nations. These countries also must provide financial support to low- and middle-income countries to handle the climate crisis. As noted in sections 3.6 and 8.1, there is both a legal and moral obligation to make this happen. Unfortunately, the wealthiest people do not seem to voluntarily change their lifestyle. Therefore, governments must take the responsibility to significantly raise taxes on consumption with high carbon footprints, from SUVs to mega yachts, private jets, space tourism, destroying the planet and letting the poorest people pay the price.

9.3 MEASURING NATIONAL EMISSIONS

There are different measures of the emissions from a country. The *national* emissions include greenhouse gases being generated within the national border, the territorial emissions. This is what is usually reported to the UN and other international bodies. From one aspect this is adequate since the nations have the responsibility for these emissions in their climate work, and they have some authority to control these emissions. Furthermore, it is relatively simple to calculate emissions from combustion within the national borders. For example, national emissions in Sweden have decreased from slightly more than 70 Mt in 1990 to 50 Mt in 2020. This is the right direction but does not show the real carbon footprint of Sweden. Also, production-based accounting does not account for carbon leakage – the phenomenon of countries reducing their domestic emissions by shifting carbon-intensive production abroad.[324]

> Burning wood
> is worsening
> climate change

Wood burning is seldom reported as part of the national emissions since it is defined as carbon emission neutral, so called biogen CO_2. One CO_2 molecule from wood has the same carbon footprint as one molecule from oil. The misinterpretation is dangerous since the CO_2 is not captured by new trees until decades later. But the climate crisis is

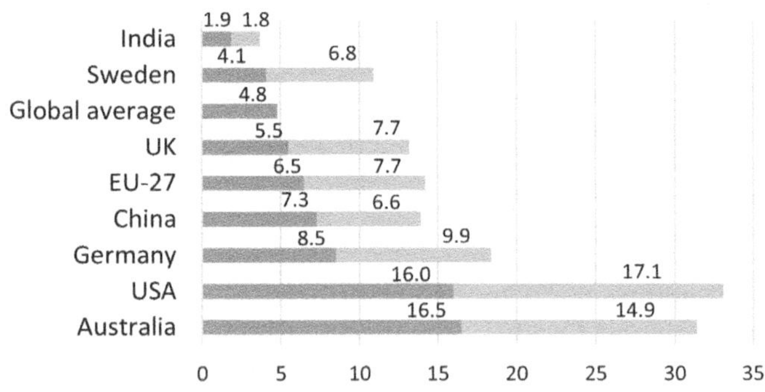

Figure 9.2 Emissions within the national borders of some countries (tons/person/year). Dark grey shows production emissions; light grey shows consumption emissions. *Source*: Our world in data & Global Carbon Project.

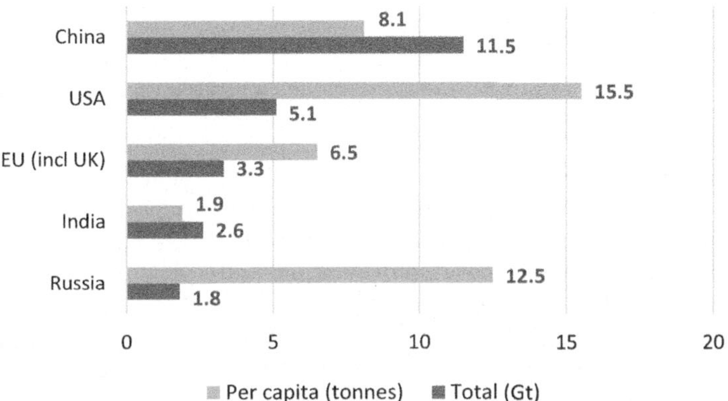

Figure 9.3 The total and per capita emissions by the world's five biggest emitters in 2019. China emits around 30% and the USA 13.5% of the global emissions. *Source*: EC Emissions database for Global Atmospheric Research.

happening now, so this type of biofuel is not a solution. If Sweden had reported biogen CO_2, then there has been no carbon footprint decrease since 1990.

> Note the difference between national and consumption emissions

The *consumption* emissions are significant and take imported and exported goods into consideration. Calculating emissions from imported goods are more uncertain. For example, to calculate emissions caused by importing a car, having components from several countries, is quite unreliable. Since there is no official statistics for consumption footprints, the numbers may vary. Still, the consumption footprint is significant. For example, total emissions for Swedish residents' air travel increased 61% between 1990 and 2014 (Figure 9.2).[325] By 2014 the air travel emissions were of similar size to Swedish road traffic. The global average carbon footprint is today around 5 tons of CO_2 per person and year and must decrease to 2.5 tons to meet the 1.5°C goal of the Paris agreement. Figure 9.2 shows that most high-income countries exceed this target by far, while low-income countries are below the targets. Figure 9.3 shows the total and per capita emissions by the world's five biggest emitters in 2019. These numbers ought to be related to the accumulated emissions, shown in Figure 3.1.

9.4 CONSUMPTION PATTERNS

The IPCC says we need to:

- Buy less meat, milk, cheese and butter.
- Eat more locally sourced seasonal food – and throw less of it away.
- Drive electric cars but walk or cycle short distances.
- Take trains and buses instead of planes.
- Use videoconferencing instead of business travel.
- Use a washing line instead of a tumble dryer.

- Insulate homes, and
- Demand low carbon in every consumer product.

We know that this is required. Let us consider a couple of consumption areas with large carbon footprint impact.

9.4.1 Apparel industry

The fashion industry accounts globally for 4–8% of all man-made greenhouse gas emissions. That's roughly the equivalent of the global aviation and shipping industries' emissions combined. The fashion industry has published information on carbon footprint reductions made in retail stores and warehouses. A major part of the apparel industry carbon footprint, however, comes from the

> Fashion industry emissions are similar to global aviation emissions.

supply chains, mostly garment manufacturing in low-income countries. They rely heavily on coal-fired energy and are also great consumers of water for raw material like cotton. Most of the manufacturing is outsourced, implying that the retail stores are only indirectly responsible for the carbon footprint of their products.

What makes the fashion industry more problematic than most other consumer goods industries is the frenetic pace of change that it encourages. Each passing season (or 'microseason') consumers are pushed to buy the latest fashion items.

The apparel industry has a large water footprint. Fabric dyes pollute water bodies, with devastating effects on aquatic life and drinking water. Trendy styles of tight jeans contain stretchy elastane synthetic material derived from plastic, which reduces recyclability and increases the environmental impact further.

Around 65% of the clothing is made of synthetic polymer polyester fabric.[326] Polyester is easy to clean and durable and is also lightweight and inexpensive. Around 70 million barrels (11 million m³) of oil a year are used to make polyester fibres in our clothes. The carbon footprint from a polyester shirt is about 5.5 kg while a cotton shirt generates 2.1 kg of CO_2.

One of the issues is PFAS (per- and polyfluoroalkyl substances), the collective name for a group of 5000 toxic (but still legal) chemicals used in many consumer products. PFAS is used to provide non-stick, water repellence, and anti-grease not only for outdoor gear but also for cosmetics, food packaging, frying pans, and firefighting foam. PFAS is not degradable in nature and is accumulated in many of the nutrient chains in nature. The industrial use of PFAS has been so high that almost every human on earth has measurable PFAS in the blood.

A good friend, Michael Greenberg, saw the direct connection between scornfulness and cotton: *'Cotton, the gift of God, requires huge amounts of water during its production. This precious fabric is washed and worn out already during the industrial process to provide jeans with fashion holes in the legs. This is nothing else than a mockery toward people that cannot afford whole and pure clothes.'* Lifestyle matters.

On the positive side, a significant effort of innovation is going into crafting lower-impact fabrics. Biocouture, fashion made from more environmentally sustainable materials, is increasingly becoming prioritized by the industry. A simple way to reduce the footprint is to buy fewer fashion clothes, swap clothes with friends, or use second-hand clothes. According to the World Bank, 40% of clothing purchased in some countries is never used.

9.4.2 Aviation

Most of the excessive carbon footprint from affluent people is related to transportation. As noted in section 6.1 some 2.8% of global CO_2 emissions come from aviation. According to the International Council on Clean Transportation (ICCT) just 3% of the global population take regular flights[327] and only 1% of the global population cause half the emissions from aviation.[328] As noted in Figure 6.9, the aviation industry emits around 1 Gt, where 85% of emissions derive from passenger transport. Fuel efficiency, measured in carbon footprint per passenger km, has improved by around 2% per year for international flights. However, while efficiency improved some 12% between 2013 and 2018, the number of passengers increased, so emissions from aviation rose by 32%. It is projected to double in 20 years. The USA, EU, and China accounted for around 55% of the CO_2 emissions from air traffic in 2019.

> 1% of the global population cause half the aviation emissions.

Emissions from air travel are significant part of the national emissions in rich countries, even if aviation contributes much less globally. For example, in Sweden international air travel has doubled in 30 years. Today emissions from air travel are as large as emissions from car transportation, and significantly higher than the steel and cement industries that are typically considered climate baddies.

Emissions from various modes of transportation are shown in Figure 6.7, while Figure 9.4 illustrates the carbon footprint per passenger km. Note that at the flight altitude, there is a larger greenhouse gas footprint than at ground.

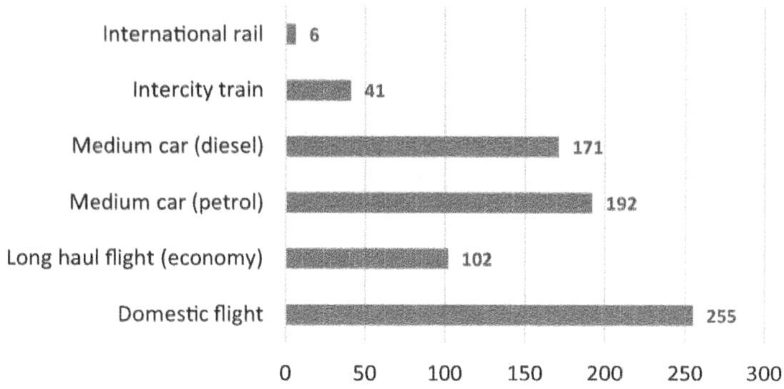

Figure 9.4 Emissions in grams of CO_2 equivalent per passenger km. The car emissions are calculated for single occupancy. The flight emissions depend on the length of the flight and the aircraft cruising altitude and weight. The domestic flights are defined as flights less than 1000 km. A long-haul flight has a lower figure on average per km because of huge emissions given off during take-off and landing.[329]

On top of the CO_2 emissions, water in the form of water vapour trails, soot, and nitrous oxides have a capability of trapping additional heat at flight altitude. The non-CO_2 emissions make up around half of the flight emissions. A passenger in premium class emits 2.6–4.3 times more CO_2 per km than a passenger in economy class.

Obviously, to travel alone in a fossil fuel car gives a carbon footprint similar to flying. An electric car getting the power from coal-powered plants will have a footprint of the same order of magnitude as a gasoline or diesel car. There is no surprise that train transportation is more eco-friendly. Often it is the attitude rather than the comfort that determines the mode of transportation. For example, going by train from my Swedish hometown of Göteborg (Gothenburg) to Stockholm (500 km) takes 3 hours by train, from city centre to city centre. Flying takes less than an hour, but usually more than 3 hours from city centre to city centre. The emissions from the flight are at least six times higher. The transportation to and from the airports has to be added.

Flying economy class London–New York roundtrip will generate of the order 1.65–1.8 tons of CO_2, depending on the calculation method used. Considering the global average footprint of 5 tons/person, this flight corresponds to one third of an annual budget.

Since the price for each transportation mode does not reflect the environmental costs, the ticket prices are often skewed. Recently I travelled to Slovenia from Sweden and took the train instead of flying. The train ticket was more expensive, but the travel was memorable. It is an enjoyable and comfortable tour through Sweden, Denmark, Germany, and Austria before entering Slovenia. The scenery was beautiful, I could read, work, relax, and eat. The travel took one day longer, but the time was used efficiently, and the memory is pleasant.

China is rapidly expanding its high-speed rail network – at the same time as building hundreds of new airports. A train trip between Beijing and Shanghai took me 5 hours of comfortable train ride at 300 km/h speed. Travelling between the city centres via the airports would have taken me longer and caused a lot more stress. The differences in carbon footprint are immense.

9.5 ACTIONS

Climate change and all its consequences cannot be solved by behavioural changes, by technical development, or political decisions in isolation. Individuals can influence decision makers and politicians, and political leaders dare to make brave decisions if citizens are prepared for it. All creative behavioural policies face one crucial condition to deliver: public acceptance. This is obvious: behavioural measures will only reach full effectiveness if citizens fully support the policy goals and instruments. If citizens perceive the scheme's impacts as socially unfair, acceptance collapses.

Anyone is looking for decent living standard with a minimum of requirements, such as:

- Shelter (housing, cooling, heating)
- Nutrition (food, food preparation, food conservation)

- Health (water, sanitation, health care)
- Socialization (communication and information, education), and
- Mobility (transport infrastructure).

The different ways to satisfy these needs are massively different in wealthy countries compared to low-income countries. We have given several examples of the huge inequalities in the world. We have to start with ourselves and our individual behaviour: the way we travel, how we consume food, clothes and gadgets, how we recycle, how we heat or cool our homes. Do we buy energy efficient equipment? Do we invest our money in sustainable industries? Any single lifestyle change will not change the world, but we have to start to change ourselves. Surely, the poorest people do not have to make further sacrifices than they already have. However, lifestyle changes are especially important in richer parts of the world where energy intensive lifestyles are the norm (and the reader of this book almost surely belongs to this group).

Technology alone is not enough: net zero emissions in 2050 cannot happen without the consent and active support of people. In part, this involves a mixture of low carbon technologies and people's engagement.

Policy makers and politicians dare to make more brave decisions if the citizens are willing to adapt. There is a true pedagogical task to explain how behavioural changes are essential to meet climate goals, to emphasize their many co-benefits, including improved health and well-being, cleaner air, less traffic congestion, cheaper fuel bills, and even fewer road traffic injuries.

We may repeat some actions that have been suggested to deal with our challenges. On the individual level:

- Make radical demand-side changes to reduce consumption to levels of sufficiency.
- Educate at all levels on the couplings between lifestyle, consumption, and climate.
- Empower young people with competent education and encouragement. Being able to communicate with other specialists is a crucial step to solve complex problems.

On the society and political level:

- Dare to talk about personal and national responsibilities.
- Realize the great environmental debt to low-income countries and act.
- Reconsider academic curricula to address complex problems.
- Professionals caught in silos should look for courses in policy making and policy theory.
- Decision makers ought to attend some courses on systems analysis or 'control for decision makers' showing how systems thinking can be applied.
- Educate on consequences of food habits, fashion trends, and consumption patterns.
- Implement adequate carbon tax on everything.

- Decide about higher taxes on wealth.
- Invest more in more climate friendly transportation. People will only drive less if they have convenient alternatives.
- Make private use of cars in cities where municipal transportation is available, relatively more expensive and less convenient.
- Facilitate the change to public transport or active mobility with additional incentives. For example, companies could give some reward to employees going to work with a minimum carbon footprint.
- Implement mandatory standards for appliances, buildings, or vehicles, and for all private investments to become low carbon over time.
- Prohibit domestic air routes when an alternative train takes less than 3–4 hours. Apply what we learnt during the pandemic, to meet online instead of travelling, whenever it is possible.

doi: 10.2166/9781789062908_0183

Chapter 10
Crisis or hope

There is another kind of power – power as empowerment of others.
Capra and Luigi Luisi (2019).

Maybe there is no decisive dramatic event that will wake us up to a common action. Still the IPCC reports are historical and the greatest summary of scientific publications on the climate ever made. The result cannot be more apparent and should urge all of us to make a quick change. It ought to be front-page news all over the world for more than a couple of days. It should lead to enormous changes and political action plans.

Is there any hope? Most people, including political leaders in the wealthy part of the world, hope to continue our lives without any great sacrifice. We hope that somebody else will make the sacrifices, since the future may become different but in a way still the same. We wish to continue our holiday travels, our consumption, and our food habits. Václav Havel, the great Czech playwright, author, turned dissident, persecuted by the communists, imprisoned, turned president of the Czech Republic, reminded us what hope is all about: *'Hope is definitely not the same thing as optimism. It is not the conviction that something will turn out well, but the certainty that something makes sense, regardless of how it turns out.'* We have to search for meaning in the climate work, not some unrealistic optimism without sacrifice. As the Trappist monk, writer and social activist Thomas Merton (1915–1968) wrote during the nonviolence peace movement work: *'The most important thing is not to be successful, but to be faithful'.*

Is the world aware of the risks to our global commons – life on Earth, fresh air and climate, oceans, forests, ice sheets, freshwater, and other processes that keep Earth stable and resilient?[330]

Leonardo Boff links global ethics and theology with the urgent challenge of ecology.[331] Focusing on the threatened Amazon of his native Brazil, Boff traces the ties that bind the fate of the rain forests with the fate of the Indians and the poor of the land. We cannot see ourselves as separated from Earth; a lifeless planet made up of 92 elements that have built up all creatures. We are more than that. We are the children of the living Earth, the Earth that has been aware of itself. This insight is no less than a revolution.

An important guidebook in my studies for this book has been a book with the provocative title *The Good Ancestor.*[332] What did I do while the Earth was unravelling? We all the time must balance between the opposing forces of short-term and long-term thinking and our time is an age of pathological short-termism. We are witnessing the destruction of the ecological systems on which life itself depends. Our unborn children as well as our small children did not contribute to this, but they have to live with the consequences. As Pope Francis proclaimed: *'intergenerational solidarity is not optional, but a basic question of justice.'*

Lately I have met many young people while lecturing on the topics of this book. Their engagement is a true source of inspiration. Many of them truly understand how our lifestyle impacts the Earth, global resources, and global warming. They are careful what they eat, they swap clothes with friends, and they are aware of the cost of transportation.

There are several organizations playing a significant role on the fight for climate justice and a better climate future. Their engagement and competence are significant. They have important education programmes. The challenge is to educate the educators and to encourage them in their struggle. Over the years the peace movement has developed an important tradition and methodology for education and empowerment. These research-based experiences can be transferred to the climate movement, providing knowledge about climate, climate crisis, and resource utilization. The international peace movement has demonstrated how common visions of the future and empowerment can achieve not only changes but even transformation of societies.

The peace movement has developed essential experiences having strong role models like Mahatma Gandhi, and the Nobel Peace Prize winners Martin Luther King Jr, Desmond Tutu, Nelson Mandela. More recent Nobel Peace Prize winners like Wangari Maathai (Kenya), Malala Yousafzai (Pakistan), and Tawakkul Karman (Yemen) have been prominent women in their struggles for nonviolence, environment, and peacebuilding, together with so many skilled and devoted leaders and educators. And remember: the IPCC got the Nobel Peace Prize in 2007 together with Al Gore.

The peace movement not only learnt nonviolent action but also constructive programmes to demonstrate concrete alternatives. Such kind of society transformation, having unconditional respect for human rights and values, and personal responsibility, can guide people also in the struggle for a better and sustainable future.

COP26 initiated the Action for Climate Empowerment (ACE) to emphasize decisions made in Rio 1992 and in Paris 2015:[333] *'Parties shall cooperate in taking measures, as appropriate, to enhance climate change education, training, public awareness, public participation, and public access to information, recognizing the importance of these steps with respect to enhancing actions under this Agreement* (Article 12, Paris).' All members of society should be empowered to engage in climate action, through education, training, public awareness, public participation, public access to information, and international cooperation on these issues. All levels of government and all sectors of society, young and old people, have to collaborate. Sustainable lifestyles, including sustainable patterns of consumption and production, are fundamental to reducing greenhouse emissions and enhancing resilience to the inevitable effects of climate change.

An increasing number of companies are comparing their production and products with the UN Sustainability Development Goals. How to be clear about sustainability and how to change some operations to reach a goal?

Belonging to the generation that has enjoyed all the benefits of cheap energy, plenty of clean water, and sufficient nutritious food I feel a combination of guilt and responsibility for the young people. My own grandchildren are among them. Their generation will govern the world in 2050. When they look back as they are growing up, will they consider us good ancestors? Have we sought to make a better world? They will never forgive us if we gave up while there was still a possibility to meet the crisis. I hope they will admit that we have tried.

doi: 10.2166/9781789062908_0187

A1

Units

LARGE NUMBERS

The data about global consumption of water and energy need large numbers. There are two primary naming systems for large numbers. The USA and France (among others) use one system, while Germany, UK, and other European countries use the other. In the USA 1 billion is 10^9 while the British name is milliard. One trillion (10^{12}) in the USA is called one billion in many European countries (one trillion in many European countries is 10^{18}). Here we consistently define billion as 10^9.

POWER AND ENERGY

The SI (International System of Units) or metric unit of **energy** is Joule.

1 J is defined as 1 Ws (wattsecond). 1 J is the designated name for the work 1 newton metre, in other words, the force 1 newton along the length 1 metre. The basic power unit watt (W) is defined as 1 J/s.

1 J = 1 Ws (wattsecond)	1 megajoule (MJ) = 10^6 J	1 gigajoule (GJ) = 10^9 J
1 terajoule (TJ) = 10^{12} J	1 petajoule (PJ) = 10^{15} J	1 exajoule (EJ) = 10^{18} J

Kilowatt-hour (kWh) is a standard unit of electrical energy.

Since 1 kW (kilowatt) = 1000 W and 1 hour = 3600 seconds we get:

1 kWh = (10^3 W) (3600 s) = 3.6×10^6 Ws = 3.6×10^6 J = 3.6 MJ (exact).

The annual electrical energy use for a nation is typically expressed in TWh.

1 TWh = 1000 GWh = 10^6 MWh = 10^9 kWh = 10^{12} Wh

The unit toe (ton of oil equivalent) is often used to indicate large energy productions. One toe is a unit of energy defined as the amount of energy released by burning one ton of crude oil. IEA and OECD define this to be 41.87 GJ or 11.63 MWh. Note that toe should be used carefully when converting electrical units. Some reports take thermal generating unit efficiency into consideration when converting kWh to toe. With a 38% plant efficiency one toe corresponds to 16 GJ.

Power is energy per time unit, the rate of energy production or consumption.

$$1 \text{ MW (megawatt)} = 10^3 \text{ kW} = 10^6 \text{ W} \qquad 1 \text{ GW (gigawatt)} = 10^3 \text{ MW}$$

A plant with the average power capacity of 1 MW will produce $1 \times 8760 = 8760$ MWh or 8.76 GWh in a year.

In a thermal power plant, we distinguish between the electrical power (MWe) and the thermal power (MWth).

MASS AND VOLUME

1 pound (lb) $= 0.4536$ kg
1 metric ton $= 0.984$ long ton or English ton
1 Mt $= 10^6$ ton 1 Gt $= 10^9$ ton

Gas emission is often measured in Tg, where 1 Tg $= 10^{12}$ g $= 10^6$ metric tons

Natural gas is converted to barrels of oil equivalent. 1 ton of oil equivalent $\cong 1125$ m^3 of natural gas This is based on the average equivalent energy content of natural gas reserves.

doi: 10.2166/9781789062908_0189

A2

Glossary

Ammonia (NH3): A compound of nitrogen and hydrogen. It can be used directly as a fuel in direct combustion process, and in fuel cells or as a hydrogen carrier. To be a low-carbon fuel, ammonia must be produced from low-carbon hydrogen, the nitrogen separated via the Haber process, and electricity needs are met by low-carbon electricity.

Anthropogenic global warming: Overall warming of Earth's climate caused or produced by humans.

Bioenergy: Energy content in solid, liquid, and gaseous products derived from biomass feedstocks and biogas. It includes solid biomass, liquid biofuels, and biogases.

Biogas: A mixture of methane, carbon dioxide, and small quantities of other gases produced by anaerobic digestion of organic matter in an oxygen-free environment.

Carbon capture, utilisation, and storage (CCS or CCUS): The process of capturing CO_2 emissions from fuel combustion, industrial processes, or directly from the atmosphere. Captured CO_2 emissions can be stored in underground geological formations, onshore or offshore, or used as an input or feedstock to create products.

Clean energy: Includes renewables, energy efficiency, low-carbon fuels, nuclear power, battery storage and carbon capture, utilization, and storage.

Concentrating solar power (CSP): Solar thermal power/electric generation systems that collect and concentrate sunlight to produce high temperature heat to generate electricity.

Geothermal: Geothermal energy is heat derived from the sub-surface of the earth. Water and/or steam carry the geothermal energy to the surface. Depending on its characteristics, geothermal energy can be used for heating and cooling purposes or be harnessed to generate clean electricity if the temperature is adequate.

Heating degree day (HDD): A measurement designed to reflect the demand for energy needed to heat building. Heating degree days are defined relative to a base temperature. The base temperature is usually an indoor temperature (between 18°C and 20°C) which is adequate for human comfort. If the outside air temperature is lower, then there is a heating requirement. The heating requirements for a given structure are considered to be directly proportional to the number of HDD at that location. A similar measurement, cooling degree day (CDD), reflects the amount of energy used to cool a building.

Natural gas: Comprises gases occurring in deposits, whether liquefied or gaseous, consisting mainly of methane. It includes both 'non-associated' gas originating from fields producing hydrocarbons only in gaseous form and 'associated' gas produced in association with crude oil as well as methane recovered from coal mines (colliery gas).

Off-grid systems: Stand-alone systems for individual households or groups of consumers.

Offshore wind: Refers to electricity produced by wind turbines that are installed in open water, usually in the ocean.

Renewables: Includes bioenergy, geothermal, hydropower, solar photovoltaics (PV), concentrating solar power (CSP), wind and marine (tide and wave) energy for electricity and heat generation.

Solar photovoltaic (PV): Electricity produced from solar photovoltaic cells.

Variable renewable energy (VRE): Refers to technologies whose maximum output at any time depends on the availability of fluctuating renewable energy resources. VRE includes a broad array of technologies such as wind power, solar PV, run-of-river hydro, concentrating solar power (where no thermal storage is included) and marine (tidal and wave).

doi: 10.2166/9781789062908_0191

A3

Abbreviations

AGGI	Annual Greenhouse Gas Index
AMOC	Atlantic Meridional Overturning Circulation
CCS	carbon capture and storage
CCUS	carbon capture, utilization, and storage
CDD	cooling degree day
CH_4	methane
CO	carbon monoxide
CO_2	carbon dioxide
$CO_{2\text{-}eq}$	carbon-dioxide equivalent
COP	Conference of the Parties (UNFCCC)
CSP	concentrating solar power
DAC	direct air capture
ESA	European Space Agency
FAO	Food and Agriculture Organization of the United Nations
GDP	Gross domestic product
GHG	greenhouse gases
HDD	heating degree day
HFC	hydrofluorocarbons
IAEA	International Atomic Energy Agency
ICCT	International Council on Clean Transportation
IEA	International Energy Agency
IEEFA	Institute for Energy Economics and Financial Analysis, India
IMF	International Monetary Fund
IPCC	Intergovernmental Panel on Climate Change
LCOE	Levelized Cost of Electricity
LED	light-emitting diode
LSLA	large-scale land acquisitions
MIT	Massachusetts Institute of Technology
NOAA	National Oceanic & Atmospheric Administration

NO_X	nitrogen oxides
N_2O	nitrous oxide
NUE	nitrogen use efficiency
OCO	Orbiting Carbon Observatory
OAPEC	Organization of the Arab Petroleum Exporting Countries
PM2.5	fine particulate matter
PPP	polluter pays principle
PV	solar photovoltaics
SCC	social cost of carbon
SDG	Sustainable Development Goals (United Nations)
SO_2	sulfur dioxide
SPIS	Solar powered irrigation systems
TCRE	transient climate response to cumulative CO_2 emissions
UNCED	UN Conference on Environment and Development
UNDP	United Nations Development Programme
UNEP	United Nations Environment Programme
UNFCCC	United Nations Framework Convention on Climate Change
VRE	variable renewable energy
VUCA	volatile, uncertain, complex, ambiguous
WCED	World Commission on Environment and Development (Brundtland commission)
WHO	World Health Organization
WMO	World Meteorological Organization
WRRF	water resource recovery facility

doi: 10.2166/9781789062908_0193

Notes

---

1 G. Brew (2019). How oil defeated the Nazis. https://oilprice.com/contributors/Gregory-Brew

2 https://en.wikipedia.org/wiki/Minamata_disease

3 Paul Ehrlich wrote the book together with his wife Anne. However, only his name was on the cover, because his publisher said '*single-authored books get much more attention than dual-authored books*'. He later confessed that '*I was at the time stupid enough to go along with it.*' The Ehrlich book and its consequences are discussed in depth in Charles Mann (2018). The book that incited a worldwide fear of overpopulation. Smithsonian Magazine, https://www.smithsonianmag.com/innovation/book-incited-worldwide-fear-overpopulation-180967499/

4 https://ourworldindata.org/world-population-growth

5 https://en.wikipedia.org/wiki/SS_Torrey_Canyon

6 https://en.wikipedia.org/wiki/Amoco_Cadiz_oil_spill

7 President Nixon to Seafarers International Union, Washington D.C., 26 November, 1973. The Seafarers International Union contributed US$100,000 to the president's 1972 campaign. See Robert Engler (1977). The Brotherhood of Oil. Energy policy and the Public Interest. A Mentor Book, New American Library, New York (Chapter 1).

8 Details of the accident are documented in *New York Times* October 1976: www.nytimes.com/1976/10/10/archives/under-the-poison-cloud-the-toxin-that-escaped-from-an-italian.html

9 Olsson (2018).

10 Schumacher (1973).

11 *The Times Literary Supplement*, 6 October, 1995, p. 39.

12 Meadows *et al.* (1972).

13 In the 1970s Olivetti was and outstanding and progressive company. They had introduced PCs that could compete with IBM/Microsoft.

14 Forrester (1971).

15   The model is available as an application package in Open Modelica, see modelica.org

16   Olsson *et al.* (2014)

17   Olsson (2012).

18   WCED (1987). *Our Common Future. World Commission on Environment and Development*, Oxford University Press, Oxford (1987).

19   Lafferty, W.M. (Ed.) (2004). *Governance for Sustainable Development. The Challenge of Adapting Form to Function.* Elgar, Cheltenham (p. 26).

20   *Scientific American*, October 26, 2015.

21   This has been carefully documented by N. Oreskes and E.M. Conway (2012).

22   G. Piani published a detailed review and criticism of the Kyoto Protocol in a book Piani (2008a) (in Italian, but there is an unofficial English version) and an English article in Piani (2008b).

23   https://en.wikipedia.org/wiki/1975_Banqiao_Dam_failure

24   A favourite definition of the scientific method is described by the famous physicist Richard Feynman: *'The only way to have real success in science …. is to describe the evidence very carefully without regard to the way you feel it should be. If you have a theory, you must try to explain what's good about it and what's bad about it equally. In science you learn a kind of standard integrity and honesty.'*

25   The pandemic starting in 2020 has demonstrated how difficult it is for virologists to convey the message about vaccinations. Many governments have followed the hard way and made vaccinations compulsory.

26   Ichimura Y., Baker D. (2019). Acute Inhalational Injury, Reference Module in Biomedical Sciences, 2019; Eckerman, I. (2005). *The Bhopal Saga— Causes and Consequences of the World's Largest Industrial Disaster.* India: Universities Press. doi:10.13140/2.1.3457.5364. ISBN 81-7371-515-7 Ingrid Eckerman was member of the International Medical Commission on Bhopal 1994. She was *persona non grata* for many years in India.

27   Hansen (2011).

28   The Chornobyl disaster is documented in a large number of publications, but the IAEA (International Atomic Energy Agency) report 'The 1986 Chornobyl nuclear power plant accident' is recommended, see https://www.iaea.org/topics/chornobyl

29   The transcript of the speech is found at https://www.margaretthatcher.org/document/107817

30   The full UNCED report (492 pages) is available at https://undocs.org/en/A/CONF.151/26/Rev.1(vol.I)

31   The Agenda 21 documentation (351 p.) is available at https://sustainabledevelopment.un.org/content/documents/Agenda21.pdf

32   The UNFCCC report (33 p.) is available at https://unfccc.int/files/essential_background/background_publications_htmlpdf/application/pdf/conveng.pdf

33   *The Guardian* 19 June 2018.

34   Pielke (2007).

35   Hansen (2011).

36 https://en.wikipedia.org/wiki/Do%C3%B1ana_disaster
37 Kanthak, J. (2000). The Baia Mare Gold Mine Cyanide Spill: Causes, Impacts and Liability. Greenpeace. https://reliefweb.int/report/hungary/baia-mare-gold-mine-cyanide-spill-causes-impacts-and-liability
38 Olsson (2018), Chapter 2.
39 Arrhenius, S. (1896).
40 Bolin (2007).
41 Pate *et al.* (2007).
42 Olsson (2012)
43 Hoffman (2016), Chapter 4.
44 UN WWDR (2014).
45 DOE (2014).
46 Olsson (2012, 2015)
47 https://www.un.org/sg/en/node/259283
48 IEA (2021a).
49 For a long-term historical perspective, the article by Pain (2017) is a gratifying reading.
50 IPCC (2021).
51 IEA (2021a).
52 http://www.fao.org/sustainability and Zabel *et al.* (2019).
53 Oxfam https://www.oxfam.org
54 UNHCR, https://www.unhcr.org/climate-change-and-disasters.html
55 The UN World Water Development Report (UN WWDR, 2020), titled *Water and Climate Change* aims at helping water communities to tackle climate change challenges.
56 Forrester (1971).
57 Simpson *et al.* (2021).
58 Harrison *et al.* (2016).
59 Wiener (1948).
60 The concept of 'wicked problems' was proposed by Rittel-Webber (1973). McMillan-Overall (2016) further compared conventional and wicked problems as outlined in Table 5.5.
61 Morton (2013).
62 UN WWDR (2014).
63 Olsson (2015).
64 Olsson (2018), Chapter 3.
65 World Economic Forum (2020).
66 https://ourworldindata.org/contributed-most-global-co2
67 Pierrehumbert (2019).
68 https://digitallibrary.un.org/record/82555
69 Rio conference 1992: https://unfccc.int/files/essential_background/background_publications_htmlpdf/application/pdf/conveng.pdf
70 IPCC (2021).
71 Cheng *et al.* (2022).
72 Tanaka and van Houtan (2022).
73 UNEP (2021).
74 WMO (2021a).

75   NAP report: https://www.nap.edu/resource/other/dels/net-zero-emissions-by-2050/?utm_source=NASEM+News+and+Publications&utm_cam paign=259a90bc3e-Eblast_COP26_Resource_2021_11_09&utm_med ium=email&utm_term=0_96101de015-259a90bc3e-103421477& goal=0_96101de015-259a90bc3e-103421477&mc_cid=259a90bc3e&mc_ eid=836701c454#page-top

76   SEI *et al.* (2021).

77   https://unfccc.int/topics/education-youth/the-big-picture/what-is-act ion-for-climate-empowerment

78   FAO-UNEP (2020).

79   WRI (2021b).

80   BBC News 11 Feb. 2022.

81   Feng *et al.* (2021).

82   https://www.greenpeace.org/africa/en/press/13072/new-illegal-forest-concessions-in-the-drc-the-environment-minister-has-done-it-again/

83   NASA News, October 2019: https://climate.nasa.gov/news/

84   USEPA (2017).

85   Callaghan *et al.* (2021).

86   Sippel *et al.* (2020).

87   Fischer *et al.* (2021).

88   Kramer and Ware (2020), Christian Aid.

89   https://news.mongabay.com/2021/12/wildlife-death-toll-from-2020-pantanal-fires-tops-17-million-study-finds/

90   UN News and UN Office for Disaster Risk Reduction (UNDRR), 1 September, 2021. https://news.un.org/en/story/2021/09/1098662

91   UNDRR (2020).

92   Raju *et al.* (2022).

93   See further Ciavarella *et al.* (2021) and the works at the Environmental Change Institute at Oxford University, UK, lead by Prof. Friederike Otto.

94   Robine *et al.* (2008).

95   Hoag (2014).

96   Vicedo-Cabrera *et al.* (2021).

97   Raymond *et al.* (2020).

98   McCrystall *et al.* (2021).

99   Temperatures are obtained from the global ERA5 dataset, produced by the Copernicus Climate Change Service. https://www.ecmwf.int/en/forecasts/dataset/ecmwf-reanalysis-v5

100 Otto *et al.* (2018).

101 https://www.bbc.com/news/world-asia-india-43252435

102 Bowman *et al.* (2020).

103 Xu *et al.* (2020).

104 IEA: https://www.iea.org/reports/driving-down-methane-leaks-from-the-oil-and-gas-industry

105 Bloomberg: https://www.bloomberg.com/features/diversified-energy-n atural-gas-wells-methane-leaks-2021/?cmpid=BBD101921_GREEN DAILY&utm_medium=email&utm_source=newsletter&utm_ term=211019&utm_campaign=greendaily

106 Ocko *et al.*, (2021).
107 Global Energy Monitor (2021).
108 IEA (2019).
109 National Oceanographic & Atmospheric Administration (NOAA): https://
gml.noaa.gov/aggi/
110 Kampschreur *et al.* (2009) is a pioneering paper.
111 https://en.wikipedia.org/wiki/Kigali_Amendment
112 Lenton (2011) and Lenton *et al.* (2019).
113 IPCC (2018).
114 Caesar *et al. (*2021).
115 GRID (2020).
116 UNU-EHS (2021).
117 Nadeau *et al.* (2021).
118 World Bank (2021a).
119 Olsson (2015) and Olsson (2018).
120 Data from https://ourworldindata.org/explorers/water-and-sanitation
121 Data from https://ourworldindata.org/diarrheal-diseases
122 https://worldwater.io/
123 UN WWDR (2018).
124 Mekonnen & Hoekstra (2016).
125 WMO (2021b).
126 WMO (2021b).
127 Reuters 2 September 2021, https://www.reuters.com/world/middle-east
/jordans-water-crisis-deepens-climate-changes-population-grows-20
21-09-02/
128 https://www.nature.com/articles/s41545-019-0039-9
129 Olsson (2015), Chapter 10.
130 https://www.unicef.org/stories/water-and-climate-change-10-things-you-
should-know
131 WMO (2021b).
132 Tellman *et al.* (2021).
133 WRI (2021a).
134 For example, Richter and Boltz (2020).
135 Hugonnet *et al.* (2021).
136 IPCC (2019).
137 https://www.worldometers.info/water/
138 Erkens and Sutanudjaja (2015).
139 https://www.waterworks.metro.tokyo.lg.jp/eng/faq/qa-7.html
140 Michelle Sneed, California Water Science Center https://www.usgs.
gov/centers/ca-water-ls/science/land-subsidence-san-joaquin-valley?qt-
science_center_objects=0#qt-science_center_objects
141 John *et al.* (2020).
142 SDSN/FEEM (2021).
143 Chiarelli *et al.* (2022).
144 Several examples are given in Olsson (2015), Chapter 2.
145 IEA (2018).

146 Olsson (2015), Chapters 15-20.
147 Olsson (2018).
148 Chini *et al.* (2021).
149 Liu *et al.* (2016).
150 Sanders and Webber (2012).
151 Kenway *et al.* (2015).
152 Kenway *et al.* (2019).
153 The REN21 *Renewables Global Status Report* (REN21, 2020) together with IRENA *Global Renewables Outlook* (2020a) provide the world's most comprehensive reports on renewables in 2020.
154 Issues related to water availability and treatment like desalination are described in Olsson (2018).
155 wad.jrc.ec.europa.eu/irrigations
156 *Times of India*, March 22, 2018.
157 *Times of India*, September 2, 2021.
158 CEA (2016).
159 IRENA (2015).
160 Casey (2013).
161 ETEnergy World 11 June 2021 https://energy.economictimes.indiatimes.com
162 IRENA (2016).
163 Ingildsen and Olsson (2016).
164 Parravicini *et al.* (2016).
165 Campana *et al.* (2021).
166 Van Rensburg (2016).
167 Al Jazeera, https://www.aljazeera.com/features/2015/5/19/siberias-tug-of-war-over-lake-baikals-water?source=nl&utm_source=nl_esq&utm_medium=email&date=120421&utm_campaign=nl25925773&list=CP
168 Yale School of Environment 2021, https://e360.yale.edu/digest/lake-powell-could-stop-producing-hydropower-in-2023-due-to-worsening-drought
169 EIA (Energy Information Administration) Short-term energy outlook.
170 World Bank (2021a).
171 *The Guardian* (2020) discussed affordability of water.
172 Ostrom (1990).
173 Figure 5.8 is published in Olsson (2015).
174 Data in Figure 5.9 are based on UNEP (2008).
175 eThekwini Municipality (2020) provides data from Durban, South Africa.
176 https://ourworldindata.org/water-use-stress#poorer-countries-used-more-water-for-agriculture
177 Hoekstra (2003).
178 Mekonnen and Hoekstra (2012).
179 https://waterfootprint.org/
180 Capra and Luisi (2019).
181 Ingildsen and Olsson (2016).
182 Ingildsen (2020).

183 Examples of how nature has been given legal rights are described for example in https://theconversation.com. Christopher Stone's game-changing work from 1972 on the legal rights of nature are described in his book Stone (2010).

184 *The Guardian*, 2 Dec., 2021.

185 Witness from Amnesty International: Episode 5 – Bodo. September 29, 2020. https://www.amnesty.org/en/latest/news/2020/09/witness-epi sode-five-bodo/

186 BP (2021).

187 https://ourworldindata.org/

188 IEA Global Energy Review 2021 https://www.iea.org/reports/ global-energy-review-2021/co2-emissions

189 https://ourworldindata.org/grapher/co2-emissions-by-fuel-line?coun try=~OWID_WRL

190 All electrical data are from https://ourworldindata.org/electricity-mix and from BP statistical Review of World Energy 2020.

191 https://www.iea.org/reports/electricity-market-report-july-2021

192 https://ourworldindata.org/electricity-mix

193 *World Resource Institute*'s Climate Data Explorer.

194 https://ourworldindata.org/co2-emissions-from-transport

195 IEA (2020).

196 https://statisticsanddata.org/data/top-countries-by-motor-vehicle-production-1950-2020/

197 International Energy Agency (IEA), https://www.iea.org/data-and-statistics/charts/transport-sector-co2-emissions-by-mode-in-the-sustainable-development-scenario-2000-2030

198 https://ourworldindata.org/co2-emissions-from-aviation

199 worldpopulationreview.com, IEA 2021, BP statistical review 2021.

200 worldpopulationreview.com, IEA 2021, BP statistical review 2021.

201 Linnerud *et al.* (2011).

202 Eurostat: https://ec.europa.eu/eurostat/cache/metadata/en/nrg_chdd_esms.htm

203 IEA (2021a), section 3.7.

204 Orr *et al.* (2022).

205 IEA (2016).

206 Alberta Energy Regulator (2021).

207 Treaty Alliance: treatyalliance.org

208 Olsson (2015), Chapter 11.

209 Ceres (2016).

210 Woda *et al.* (2020).

211 Data from World Bank 2021.

212 IEA (2021c).

213 BP (2021).

214 Olsson (2015), Chapter 11.

215 Zabbey and Olsson (2017).

216 UNEP (2011).

217 Amnesty International (2020).

218 Lark *et al.* (2022).
219 Rodionova *et al.* (2016).
220 Njenga and Mendum (2018).
221 Olsson (2015), Chapter 12, Gerbens-Leenes *et al.* (2012), and IEA (2016).
222 Kiniry *et al.* (2008).
223 Manuel (2007).
224 Olsson (2015), Chapter 12.
225 EASAC, May 2020: https://easac.eu/media-room/press-releases/details/european-national-academies-of-science-easacs-message-to-iea-bioenergy-lets-debate-the-real-issues/
226 Lazard's Levelized Cost of Energy Analysis (LCOE 15.0), www.lazard.com/perspective/levelized-cost-of-energy-levelized-cost-of-storage-and-levelized-cost-of-hydrogen/
227 Source IRENA (2020a).
228 Olsson (2015), Chapter 10.
229 Ali and El-Magd (2016).
230 Ha and Seth (2021).
231 IEA (2021a), p 22.
232 IAEA (2021).
233 The quotation is from an open letter, signed by James Hansen, atmospheric scientist Ken Caldeira, meteorologist Kerry Emanuel and climate scientist Tom Wigley. The letter is addressed 'To those influencing environmental policy but opposed to nuclear power.' It is cited for example in *World Nuclear News*, November 2013, see https://www.world-nuclear-news.org/EE-Nuclear-essential-for-climate-stability-0411137.html
234 EU taxonomy info: https://eu-taxonomy.info/news/european-parliament-adopts-the-taxonomy-regulation
235 Olsson (2015), Chapter 13.
236 Realmonte *et al.* (2019).
237 Conca J. (2019).
238 Lyons (2021).
239 IEA (2021d).
240 Sir David Attenborough speaking at the G7 summit hosted by the UK in Cornwall in the summer of 2021.
241 Skyllas-Kazacos and McCann (2015).
242 Shiva Kumar and Himabindu (2019) present a good review.
243 IRENA (2020b).
244 UNCTAD (2020) and https://unctad.org/news/developing-countries-pay-environmental-cost-electric-car-batteries
245 IEA (2021e).
246 IEA (2021b), section 6.3.
247 IEA (2021e).
248 Niarchos (2021).
249 Alex Crawford, Sky News, 2017, https://news.sky.com/story/meet-dorsen-8-who-mines-cobalt-to-make-your-smartphone-work-10784120
250 Amnesty International (2016).

251 Melville (2020).
252 Hochschild (1998).
253 Van Brusselen *et al.* (2020).
254 Lawson (2021).
255 www.statista.com/statistics/268789/countries-with-the-largest-produc tion-output-of-lithium/
256 BBC 2021: https://www.bbc.com/future/article/20210809-how-your-phone-battery-creates-striking-landscapes
257 Åström *et al.* (2021).
258 Ourworldindata.org
259 Hendryx *et al.* (2020).
260 Laumbach and Kipen (2012).
261 FAO measures food insecurity using the Food Insecurity Experience Scale (FIES).
262 FAO *et al.* (2021).
263 Holleman *et al.* (2020).
264 FAO (2021).
265 Poore and Nemecek (2018) and error corrections Poore and Nemecek (2021).
266 Lappé (1971).
267 Buchman (1973).
268 https://ourworldindata.org/meat-production#which-countries-eat-the-most-meat
269 Source FAO (2017).
270 https://www.theworldcounts.com/challenges/consumption/foods-and-beverages/world-consumption-of-meat/story
271 WWF report: https://ecobnb.com/blog/2018/01/meat-consumption-rep ort-wwf/
272 Data from https://ourworldindata.org/land-use
273 Katherine Latham, BBC, February 2022, https://www.bbc.com/future/bespoke/follow-the-food/how-to-use-food-waste-for-good/
274 UNEP (2022).
275 FAO (2020).
276 The *National Geographic* magazine had an extensive description of the Aral Sea catastrophe in the 2 Oct., 2014 issue.
277 Zhu *et al.* (2019).
278 Klapwijk *et al.* (2014).
279 Olsson (2018).
280 FAO (2018).
281 Smil (2004).
282 https://ourworldindata.org/fertilizers
283 International Fertilizer Organization (IFA), https://www.fertilizer.org/
284 UNIDO (2014).
285 Pollan (2007).
286 Montgomery (2017).
287 Dr Elaine Ingham, a leader in soil microbiology emphasizes the importance of the micro-life in the soil, see for example https://www.soilfoodweb.com/

288 Six *et al.* (2000).
289 Savory (2016).
290 Von Bremen and Rundgren (2020).
291 The Netflix documentary 'Kiss the ground' (2020) examines how a healthy food production is a decisive factor for our climate and for our future on Earth.
292 Presentation by FAO Director-General José Graziano da Silva in 2019, https://www.fao.org/news/story/en/item/1203200/icode/.
293 EAT-Lancet Commission (2021).
294 Chattopadhyay (2014).
295 http://paris-equity-check.org/the-science.html
296 https://unfccc.int/process-and-meetings/the-paris-agreement/nationally-determined-contributions-ndcs/nationally-determined-contributions-ndcs
297 Stiglitz (2013).
298 FAO *et al.* (2021).
299 Diamond (2005).
300 Meadows *et al.* (1972).
301 Schumacher (1973).
302 https://www.forbes.com/billionaires/
303 International Monetary Fund (2021).
304 Climate Vulnerable Forum https://thecvf.org/
305 Batten (2018).
306 Nordhaus (2013).
307 Revesz *et al.* (2014).
308 Rio Declaration 1992: https://www.un.org/en/development/desa/population/migration/generalassembly/docs/globalcompact/A_CONF.151_26_Vol.I_Declaration.pdf
309 Stern (2006).
310 Stern *et al.* (2022).
311 World Bank (2021b).
312 Bloomberg (2021).
313 World Economic Forum, April 2021. https://www.weforum.org/agenda/2021/04/economists-global-action-climate-change-natural-disasters
314 Raworth (2017).
315 Rockström *et al.* (2009).
316 Duncan (1975).
317 Raworth (2017), Chapter 6.
318 www.oecd.org/environment/
319 IEA (2021b).
320 Bloomberg (2021).
321 IISD (International Institute for Sustainable Development), Global Subsidies Initiative, http://www.iisd.org/gsi/subsidy-watch-blog/53-ways-reform-fossil-fuel-consumer-subsidies-and-pricing
322 Oxfam (2022).
323 Oxfam (2021).

324 Kander *et al.* (2015).
325 Larsson *et al.* (2018).
326 Dr. Lynn Wilson, University of Glasgow, UK. Fashion consumer behaviour: https://www.circulareconomywardrobe.co.uk/
327 ICCT: https://theicct.org/publications/co2-emissions-commercial-aviation-2020
328 Gössling and Humpe (2020).
329 https://www.statista.com/statistics/
330 Gaffney *et al.* (2021).
331 Boff (1997).
332 Krznaric (2020).
333 https://unfccc.int/topics/education-youth/the-big-picture/what-is-action-for-climate-empowerment

doi: 10.2166/9781789062908_0205

References

Note: latest access to all websites March 2022

Alberta Energy Regulator (2021). Bird protection from tailings ponds in the minable oilsands: Review of current approaches, knowledge gaps, and recommendations for better practice. Report. https://www.scribd.com/document/489963319/Bird-protection-from-tailings-ponds-in-the-minable-oilsands-Review-of-current-approaches-knowledge-gaps-and-recommendations-for-better-practice

Ali E. M. and El-Magd I. A. (2016). Impact of human interventions and coastal processes along the Nile Delta coast, Egypt during the past twenty-five years. *The Egyptian Journal of Aquatic Research*, **42**(1), 1–10, https://doi.org/10.1016/j.ejar.2016.01.002

Amnesty International (2016). 'This is What we Die For'. Human Rights Abuses in the Democratic Republic of the Congo Power the Global Trade in Cobalt. Amnesty International Ltd, London, UK.

Amnesty International (2020). No cleanup, no justice: Shell's oil pollution in the Niger Delta. https://www.amnesty.org

Arrhenius S. (1896). On the influence of carbonic acid in the air upon the temperature of the ground. Philosophical Magazine and Journal of Science (5th series, London, Edinburgh, and Dublin), **41**, 237–276, https://doi.org/10.1080/14786449608620846

Åström S., Källmark L., Yaramenka K. and Grennfelt P. (2021). European and Central Asian Actions on Air Quality, IVL Swedish Environmental Research Institute, Report C598, https://www.ivl.se/

Batten S. (2018). Climate change and the macro-economy: a critical review. Staff working paper 706, Bank of England, SSRN eLibrary, https://papers.ssrn.com/sol3/papers.cfm?abstract_id=3104554

Bloomberg (2021). Climate policy factbook. Bloomberg NEF, July 20, 2021. https://www.bloomberg.org/publications/

Boff L. (1997). Cry of the Earth, cry of the Poor. Orbis Series of Global Ecology, USA.

Bolin B. (2007). A History of the Science and Politics of Climate Change. Cambridge University Press, Cambridge, UK.

Bowman D. M. J. S., Kolden C. A., Abatzoglou J. T., Johnston F. H., van der Werf G. R. and Flannigan M. (2020). Vegetation fires in the anthropocene. *Nature Reviews Earth & Environment*, **1**, 500–515, https://doi.org/10.1038/s43017-020-0085-3

BP (2021). Statistical Review of World Energy, 70th edn. https://www.bp.com/content/dam/bp/business-sites/en/global/corporate/pdfs/energy-economics/statistical-review/bp-stats-review-2021-full-report.pdf

BP Energy outlook (2014). BP energy outlook 2035. bp.com/energyoutlook

Buchman E. E. (1973). Recipes for A Small Planet. Ballantine Books, New York.

Caesar L., McCarthy G. D., Thornalley D. J. R., Cahill N. and Rahmstorf S. (2021). Current atlantic meridional overturning circulation weakest in last millennium. *Nature Geoscience*, **14**, 118–120, https://www.nature.com/articles/s41561-021-00699-z https://doi.org/10.1038/s41561-021-00699-z

Callaghan M., Schleussner C.-F., Nath S., Lejeune Q., Knutson T., Reichstein M., Hansen G., Theokritoff E., Andrijevic M., Brecha R. J., Hegarty M., Jones C., Lee K., Lucas A., Maanen N., Menke I., Pfleiderer P., Yesil B. and Minx J. (2021). Machine-learning-based evidence and attribution mapping of 100,000 climate impact studies. *Nature Climate Change*, **11**, 966–972. https://doi.org/10.1038/s41558-021-01168-6

Campana P. E., Mainardis M., Moretti A. and Cottes M. (2021). 100% Renewable wastewater treatment plants: techno-economic assessment using a modelling and optimization approach. *Energy Conversion and Management*, **239**, 114214, https://doi.org/10.1016/j.enconman.2021.114214

Capra F. and Luigi Luisi P. (2019). The Systems View of Life. A Unifying Version. 10th printing, Cambridge University Press, UK.

Casey A. (2013). Reforming Energy Subsidies Could Curb India's Water Stress, http://www.worldwatch.org/

CEA (2016). Growth of Electricity Sector in India From 1947–2013. Central Electricity Authority, New Delhi, India, 2013. www.indiaenvironmentportal.org.in/content/380722/growth-of-electricity-sector-in-india-from-1947-2013/

Ceres (2016). Hydraulic fracturing & water stress: Water demand by the numbers. Ceres report. https://www.ceres.org/resources/reports/hydraulic-fracturing-water-stress-water-demand-numbers

Chattopadhyay M. K. (2014). Use of antibiotics as feed additives: A burning question. *Frontiers in Microbiology*, **5**, 1–3, https://doi.org/10.3389/fmicb.2014.00334

Cheng L., Abraham J., Trenberth K. E., *et al.* (2022). Another record: ocean warming continues through 2021 despite La Niña conditions. *Advances in Atmospheric Sciences*, **39**, 373–385.

Chiarelli D. D., D'Odorico P., Müller M. F., *et al.* (2022). Competition for water induced by transnational land acquisitions for agriculture. *Nature Communications*, **13**, 505, https://doi.org/10.1038/s41467-022-28077-2

Chini C. M., Excell L. E. and Stillwell A. S. (2021). A review of energy-for-water data in energy-water nexus publications. *Environmental Research Letters*, **15**(12), 123011, https://doi.org/10.1088/1748-9326/abcc2a

Ciavarella A., Cotterill D., Stott P., Kew S., Philip S., van Oldenborgh G. J., Skålevåg A., Lorenz P., Robin Y., Otto F., Hauser M., Seneviratne S. I., Lehner F. and Zolina O. (2021). Prolonged Siberian heat of 2020 almost impossible without human influence. *Climatic Change*, **166**(9), 1–18.

Conca J. (2019). Extract CO_2 from our air, use it to create synthetic fuels. Energypost.eu, October 11, 2019.

Diamond J. (2005). Collapse: How Societies Choose to Fail or Survive. Penguin, London.

DOE (2014). The Water-Energy Nexus: Challenges and Opportunities. U.S. Department of Energy, Washington DC, USA, https://www.energy.gov/sites/default/files/2014/07/f17/Water%20Energy%20Nexus%20Full%20Report%20July%202014.pdf

Duncan O. D. (1975). Does money buy satisfaction? *Social Indicators Research*, **2**(3), 267–274. https://www.jstor.org/stable/27521760 https://doi.org/10.1007/BF00293248

EAT-Lancet Commission (2021). Food Planet Health: the EAT-Lancet Commission on Healthy Diets from Sustainable Food Systems (W. Willett and J. Rockström, co-chairs). https://eatforum.org/content/uploads/2019/07/EAT-Lancet_Commission_Summary_Report.pdf

Erkens G. and Sutanudjaja E. H. (2015). Towards a global land subsidence map. *Proceedings of the International Association of Hydrological Sciences*, **372**, 83–87. See https://www.researchgate.net/publication/https://doi.org/10.5194/piahs-372-83-2015

eThekwini Municipality (2020). http://www.durban.gov.za/City_Services/water_sanitation/Pages/default.aspx

FAO (2017). Guidelines for the Compilation of Food Balance Sheets. Food and Agriculture Organization of the United Nations, Rome, www.fao.org

FAO (2018). The Benefits and Risks of Solar-Powered Irrigation–A Global Overview, Food and Agriculture Organization of the United Nations, Rome, www.fao.org

FAO (2020). The State of Food and Agriculture 2020. Overcoming Water Challenges in Agriculture. Food and Agriculture Organization of the United Nations, Rome, https://doi.org/10.4060/cb1447en

FAO (2021). The Impact of Disasters and Crises on Agriculture and Food Security. Food and Agriculture Organization of the United Nations, Rome. www.fao.org/3/cb3673en/cb3673en.pdf

FAO, IFAD, UNICEF, WFP, WHO (2021). The State of Food Security and Nutrition in the World 2021. Transforming Food Systems for Food Security, Improved Nutrition and Affordable Healthy Diets for all. FAO, Rome, https://doi.org/10.4060/cb4474en

FAO, UNEP (2020). The State of the World's Forests 2020. Forests, Biodiversity and People. Food and Agriculture Organization of the United Nations, Rome, https://doi.org/10.4060/ca8642en

Feng X., Merow C., Liu Z., *et al.* (2021). How deregulation, drought and increasing fire impact Amazonian biodiversity. *Nature*, **597**, 516–521, https://doi.org/10.1038/s41586-021-03876-7

Fischer E. M., Sippel S. and Knutti R. (2021). Increasing probability of record-shattering climate extremes. *Nature Climate Change*, **11**, 689–695, https://doi.org/10.1038/s41558-021-01092-9

Forrester J. W. (1971). World Dynamics. WrightAllen Press, Cambridge, Mass.

Gaffney O., Tcholak-Antitch Z., *et al.* (2021). Global Commons Survey: Attitudes to planetary stewardship and transformation among G20 countries. Global Commons Alliance, globalcommonsalliance.org

Gerbens-Leenes P. W., van Lienden A. R., Hoekstra A. Y. and van der Meer Th. H. (2012). Biofuel scenarios in a water perspective: The global blue and green water footprint of road transport in 2030. *Global Environmental Change*, **22**, 764–775, https://doi.org/10.1016/j.gloenvcha.2012.04.001

Global Energy Monitor (2021). Coal Mine Methane on the Brink. https://globalenergymonitor.org/report/coal-mine-methane-on-the-brink/

Gössling S. and Humpe A. (2020). The global scale, distribution and growth of aviation: implications for climate change. *Global Environmental Change*, **65**, 102194, https://doi.org/10.1016/j.gloenvcha.2020.102194

GRID (2020). Global report on Internal Displacement. IDMC (Internal Displacement Monitoring Centre) and NRC (Norwegian Refugee Council). https://www.internal-displacement.org/sites/default/files/publications/documents/2020-IDMC-GRID.pdf

Guardian (2020). The Affordability of Water and Wastewater Service in Twelve U.S. Cities: A Social, Business, and Environmental Concern. https://www.theguardian.com/environment/2020/jun/23/full-report-read-in-depth-water-poverty-investigation

Ha H. T. and Seth F. N. (2021). The Mekong River Ecosystem in Crisis: ASEAN Cannot be a Bystander. ASEAN Studies Centre, ISEAS – Yusof Ishak Institute. https://www.iseas.edu.sg/articles-commentaries/iseas-perspective/2021-69-the-mekong-river-ecosystem-in-crisis-asean-cannot-be-a-bystander-by-hoang-thi-ha-and-farah-nadine-seth/

Hansen J. (2011). Storms of my Grandchildren: The Truth About the Coming Climate Catastrophe and our Last Chance to Save Humanity. Bloomsbury, London, New Delhi, New York, Sydney.

Harrison P. A., Dunford R. W., Holman I. P. and Rounsevell M. D. A. (2016). Climate change impact modelling needs to include cross-sectoral interactions. *Nature Climate Change*, **6**, 885–890, https://doi.org/10.1038/nclimate3039

Hendryx M., Zullig K. J. and Luo J. (2020). Impacts of coal use on health. *Annual Review of Public Health*, **41**, 397–415. https://www.annualreviews.org/doi/10.1146/annurev-publhealth-040119-094104 https://doi.org/10.1146/annurev-publhealth-040119-094104

Hoag H. (2014). Russian summer tops 'universal' heatwave index. Nature. https://www.nature.com/news/russian-summer-tops-universal-heatwave-index-1.16250

Hochschild A. (1998). King Leopold's Ghost. Mariner Books, USA.

Hoekstra A. Y. (ed.) (2003). Virtual Water Trade. Proceedings of the International Experts Meeting on Virtual Water Trade. Value of Water Research Report Series No. 12, IHE, Delft, www.waterfootprint.org/Reports/Report12.pdf

Hoffman A. R. (2016). The U.S. Government & Renewable Energy: A Winding Road. Pan Stanford Publishing Pte. Ltd., Singapore.

Holleman C., Rembold F., Cresp O. and Conti V. (2020). The Impact of Climate Variability and Extremes on Agriculture and Food Security–An Analysis of the Evidence and Case Studies. Background paper for The State of Food Security and Nutrition in the World 2018. FAO, Rome, https://doi.org/10.4060/cb2415en

Howard G., Calow R., Macdonald A. and Bartram J. (2016). Climate change and water and sanitation: likely impacts and emerging trends for action. *Annual Review of Environment and Resources*, **41**, 253–276, https://doi.org/10.1146/annurev-environ-110615-085856

Hugonnet R., McNabb R., Berthier E., *et al.* (2021). Accelerated global glacier mass loss in the early twenty-first century. *Nature*, **592**, 726–731, https://doi.org/10.1038/s41586-021-03436-z

IAEA (2021). Nuclear power reactors in the world. International Atomic Energy Agency. Reference data series No. 2/41.

IEA (2012). World Energy Outlook. International Energy Agency, Paris, France. https://www.iea.org/reports/world-energy-outlook

IEA (2016). Water Energy Nexus. Exerpt from the World Energy Outlook 2016. https://www.iea.org/reports/world-energy-outlook-2016

IEA (2018). World Energy Outlook. International Energy Agency, Paris. https://www.iea.org/reports/world-energy-outlook-2018

IEA (2019). World Energy Outlook. International Energy Agency, Paris. https://www.iea.org/reports/world-energy-outlook-2019

IEA (2020). Energy Technology Perspectives 2020, International Energy Agency, Paris. https://www.iea.org

IEA (2021a). World Energy Outlook. International Energy Agency, Paris. https://www.iea.org/reports/world-energy-outlook-2021

IEA (2021b). Net Zero by 2050. A Roadmap for the Global Energy Sector. International Energy Agency, Paris. www.iea.org/reports/

IEA (2021c). Flaring emissions. Tracking Report, November 2021. International Energy Agency, www.iea.org/reports/

IEA (2021d). CCUS around the world. Technology Report, April 2021. International Energy Agency. https://www.iea.org/reports

IEA (2021e). The role of critical minerals in clean energy transitions. Part of World Energy outlook 2021 (IEA, 2021a).

Ingildsen P. (2020). Water Stewardship. IWA Publishing, London. Open access available at: https://iwaponline.com/ebooks/book-pdf/701561/wio9781789060331.pdf

Ingildsen P. and Olsson G. (2016). Smart Water Utilities. IWA Publications, London. Open Access 2020: https://iwaponline.com/ebooks/book/11/Smart-Water-Utilities-Complexity-Made-Simple

International Monetary Fund (2021). Fiscal Monitor April 2021: A Fair Shot. https://www.imf.org/en/Publications/

IPCC (2018). Global warming of 1.5°C. In: An IPCC Special Report on the Impacts of Global Warming of 1.5°C Above pre-Industrial Levels and Related Global Greenhouse gas Emission Pathways, in the Context of Strengthening the Global Response to the Threat of Climate Change, V. Masson-Delmotte, P. Zhai, H. O. Pörtner, D. Roberts, J. Skea, P. R. Shukla, A. Pirani, W. Moufouma-Okia, C. Péan, R. Pidcock, S. Connors, J. B. R. Matthews, Y. Chen, X. Zhou, M. I. Gomis, E. Lonnoy, T. Maycock, M. Tignor and T. Waterfield (eds.).

IPCC (2019). In: IPCC Special Report on the Ocean and Cryosphere in A Changing Climate, H.-O. Pörtner, D. C. Roberts, V. Masson-Delmotte, P. Zhai, M. Tignor, E. Poloczanska, K. Mintenbeck, A. Alegría, M. Nicolai, A. Okem, J. Petzold, B. Rama and N. M. Weyer (eds.), Cambridge University Press, Cambridge, UK and New York, USA.

IPCC (2021). Climate change 2021: The physical science basis. In: Contribution of Working Group I to the Sixth Assessment Report of the Intergovernmental Panel on Climate Change, V. Masson-Delmotte, P. Zhai, A. Pirani, S. L. Connors, C. Péan, S. Berger, N. Caud, Y. Chen, L. Goldfarb, M. I. Gomis, M. Huang, K. Leitzell, E. Lonnoy, J. B. R. Matthews, T. K. Maycock, T. Waterfield, O. Yelekçi, R. Yu and B. Zhou (eds.), Cambridge University Press, Cambridge, UK and New York, USA.

IRENA (2015). Renewable Energy in the Water, Energy and Food Nexus, IRENA, Abu Dhabi. www.irena.org/

IRENA (2016). Solar Pumping for Irrigation: Improving Livelihoods and Sustainability, The International Renewable Energy Agency, Abu Dhabi. https://www.irena.org/

IRENA (2020a). Global Renewables Outlook. International Renewable Energy Agency, Abu Dhabi. www.irena.org/

IRENA (2020b). Green Hydrogen. A Guide to Policy Making. International Renewable Energy Agency, Abu Dhabi, ISBN: 978-92-9260-286-4, www.irena.org/publications

John A., Nathan R., Horne A., Stewardson M. and Webb J. A. (2020). How to incorporate climate change into modelling environmental water outcomes: A review. *Journal of Water and Climate Change*, **11**(2), 327–340, https://doi.org/10.2166/wcc.2020.263

Kampschreur M. J., Temmink H., Kleerebezem R., Jetten M. S. M. and van Loosdrecht M. C. M. (2009). Nitrous oxide emission during wastewater treatment. *Water Research*, **43**, 4093–4103, https://doi.org/10.1016/j.watres.2009.03.001

Kander A., Jiborn M., Moran D. and Wiedmann T. (2015). National greenhouse-gas accounting for effective climate policy on international trade. *Nature Climate Change*, **5**(5), 431–435, https://doi.org/10.1038/nclimate2555

Kenway S. J., Binks A., Lane J., Lant P. A., Lam K. L. and Simms A. (2015). A systemic framework and analysis of urban water energy. *Environmental Modelling & Software*, **73**, 272–285, https://doi.org/10.1016/j.envsoft.2015.08.009

Kenway S. J., Lam K. L., Stokes-Draut J., Twomey Sanders K., Binks A. N., Bors J., Head B., Olsson G. and McMahon J. E. (2019). Defining water-related energy for global comparison, clearer communication, and sharper policy. *Journal of Cleaner Environment*, **236**, 1–13, https://www.sciencedirect.com/science/article/pii/S0959 652619323108?via%3Dihub

Kiniry J. R., Lynd L., Greene N., Johnson M-V. V., Casler M. and Laser M. S. (2008). Biofuels and water use: comparison of maize and switchgrass and general perspectives. Chapter 2 in: New Research on Biofuels, J. H. Wright and D. A. Evans (eds.), Nova Science Publishers, Inc.

Klapwijk C. J., van Wijk M. T., Rosenstock T. S., van Asten P. J. A., Thornton P. K. and Giller K. E. (2014). Analysis of trade-offs in agricultural systems: current status and way forward. *Current Opinion in Environmental Sustainability*, **6**, 110–115, ISSN 1877-3435, https://www.sciencedirect.com/science/article/pii/ S1877343513001607 https://doi.org/10.1016/j.cosust.2013.11.012

Kramer K. and Ware J. (2020). Counting the cost 2020: A year of climate breakdown. Christian Aid Report. christianaid.org.uk

Krznaric R. (2020). The Good Ancestor. How to Think Long Term in A Short-Term World. Penguin, Random House, UK.

Lafferty W. M. (ed.) (2004). Governance for Sustainable Development. The Challenge of Adapting Form to Function. Elgar, Cheltenham, p. 26.

Lappé F. M. (1971). Diet for a Small Planet. Ballantine Books, New York, USA.

Lark T. J., Hendricks N. P., Smith A., Patese N., Spawn-Lee S. A. Bougiea M., Booth E. G., Kucharika C. J. and Gibbs H. K. (2022). Environmental outcomes of the US renewable fuel standard. *Proceedings of the National Academy of Sciences*, **119**(9), 1–8, https://doi.org/10.1073/pnas.2101084119

Larsson J., Kamb A., Nässén J. and Åkerman J. (2018). Measuring greenhouse gas emissions from international air travel of a country's residents' methodological development and application for Sweden. *Environmental Impact Assessment Review*, **72**, 137–144, https://doi.org/10.1016/j.eiar.2018.05.013

Laumbach R. J. and Kipen H. M. (2012). Respiratory health effects of air pollution: update on biomass smoke and traffic pollution. *Journal of Allergy and Clinical Immunology*, **129**(1), 3–11, https://doi.org/10.1016/j.jaci.2011.11.021

Lawson M. F. (2021). The DRC mining industry. Child labor and formalization of small-scale mining. Wilson Center. https://www.wilsoncenter.org/blog-post/ drc-mining-industry-child-labor-and-formalization-small-scale-mining

Lenton T. (2011). Early warning of climate tipping points. *Nature Climate Change*, **1**, 201–209, https://doi.org/10.1038/nclimate1143

Lenton T. M., Rockström J., Gaffney O., Rahmstorf S., Richardson K., Steffen W. and Schnellhuber H. J. (2019). Climate tipping points – too risky to bet against. *Nature*, **575**, 592–595, https://doi.org/10.1038/d41586-019-03595-0

Linnerud K., Mideksa T. K. and Eskeland G. S. (2011). The impact of climate change on nuclear power supply. *The Energy Journal*, **32**(1), 149–. https://www.iaee. org/energyjournal/article/2411 https://doi.org/10.5547/ISSN0195-6574-EJ -Vol32-No1-6

Liu Y., Hejazi M., Kyle P., Ki S. H., Davies E., Miralles D. G., Teuling A. J., He Y. and Niyogi D. (2016). Global and regional evaluation of energy for water. *Environmental Science and Technology*, **50**, 9736–9745, https://doi.org/10.1021/acs.est.6b01065

Lyons M. (2021). Renewed interest in Carbon Capture strategies for net-zero: targets, obstacles, costs, priorities. Energypost.eu, November 10.

Manuel J. (2007). Battle of the biofuels. *Environmental Health Perspectives*, **115**(2), A92–A95, https://doi.org/10.1289/ehp.115-a92

McCrystall M. R., Stroeve J., Serreze M., Forbes B. C. and Screen J. A. (2021). New climate models reveal faster and larger increases in Arctic precipitation than previously projected. *Nature Communications*, **12**, 6765, https://doi.org/10.1038/s41467-021-27031-y

McMillan C. and Overall J. (2016). Wicked problems: turning strategic management upside down. *Journal of Business Strategy*, **37**(1), 34–43, https://doi.org/10.1108/JBS-11-2014-0129

Meadows D. H., Meadows D. L., Randers J. and Behrens III W. W. (1972). The Limits to Growth. New American Library, New York, USA.

Mekonnen M. M. and Hoekstra A. Y. (2012). A global assessment of the water footprint of farm animal products. *Ecosystems*, **15**(3), 401–415, https://doi.org/10.1007/s10021-011-9517-8

Mekonnen M. M. and Hoekstra A. Y. (2016). Four billion people facing severe water scarcity. *Science Advances*, **2**, e1500323, https://doi.org/10.1126/sciadv.1500323

Melville J. (2020). From stone to phone. Modern day cobalt slavery in Congo. Byline Times, https://bylinetimes.com/2020/06/19/from-stone-to-phone-modern-day-cobalt-slavery-in-congo/

Montgomery D. (2017). Growing A Revolution: Bringing Our Soil Back to Life. W.W. Norton & Company, New York, London.

Morton T. (2013). Hyperobjects. Philosophy and Ecology After the End of the World. University of Minnesota Press, Minneapolis, MN, USA.

Nadeau K. C., Agache I., Jutel M., Annesi Maesano I., Akdis M., Sampath V., D'Amato G., Cecchi L., Traidl-Hoffmann C. and Akdis C. A. (2021). Climate change: A call to action for the united nations. Allergy, **77**, 1087–1090, https://doi.org/10.1111/all.15079

Niarchos N. (2021). The dark side of Congo's cobalt rush. The New Yorker, 21 May 2021. https://www.newyorker.com/magazine/2021/05/31/the-dark-side-of-congos-cobalt-rush

Njenga M. and Mendum R. (eds.) (2018). Recovering Bioenergy in Sub-Saharan Africa: Gender Dimensions, Lessons and Challenges. International Water Management Institute (IWMI), Colombo, Sri Lanka. iwmi.cgiar.org/publications

Nordhaus W. D. (2013). The Climate Casino: Risk, Uncertainty, and Economics for A Warming World. Yale University Press, New Haven, CT.

Ocko I. B., Sun T., Shindell D., Oppenheimer M., Hristov A. N., Pacala S. W., Mauzerall D. L., Xu Y. and Hamburg S. P. (2021). Acting rapidly to deploy readily available methane mitigation measures by sector can immediately slow global warming. *Environmental Research Letters*, **16**(5), 4042.

Olsson G. (2012). ICA and me – a subjective review. *Water Research*, **46**(6), 1585–1624, https://doi.org/10.1016/j.watres.2011.12.054

Olsson G. (2015). Water and Energy–Threats and Opportunities, 2nd edn. IWA Publications, London.

Olsson G. (2018). Clean Water Using Solar and Wind: Outside the Power Grid. IWA Publications, London. Open Access 2019: https://iwaponline.com/ebooks/book/738/Clean-Water-Using-Solar-and-Wind-Outside-the-Power

Olsson G., Carlsson B., Comas J., Copp J., Gernaey K. V., Ingildsen P., Jeppsson U., Kim C., Rieger L., Rodríguez-Roda I., Steyer J.-P., Takács I., Vanrolleghem P. A., Vargas Casillas A., Yuan Z. and Åmand L. (2014). Instrumentation, control and automation in wastewater–from London 1973 to narbonne 2013. *Water Science and Technology*, **69**(7), 1373–1385, https://doi.org/10.2166/wst.2014.057

Oreskes N. and Conway E. M. (2012). Merchants of Doubt. Bloomsbury Publishing, London, UK.

Orr M., Dingle G., Kellison T. and Ross W. J. (2022). Slippery slopes. How climate change is threatening the winter Olympics. Report produced by Loughborough

University, London. https://www.sportecology.org/_files/ugd/a700be_9aa3ec697
a39446eb11b8330aec19e30.pdf?utm_source=braze&utm_medium=email&utm_
campaign=newsletter-columnist-2022-02-04

Ostrom E. (1990). Governing the Commons: The Evolution of Institutions for Collective
Action. Cambridge University Press, New York, USA.

Otto F. E. L., Wolski P., Lehner F., Tebaldi C., van Oldenborgh G. J., Hogesteeger S., Singh
R., Holden P., Fučkar N. S., Odoulami R. C. and New M. (2018). Anthropogenic
influence on the drivers of the western cape drought 2015–2017. *Environmental
Research Letters*, **13**, 124010, https://doi.org/10.1088/1748-9326/aae9f9

Oxfam (2021). Carbon inequality in 2030. https://www.oxfam.org/en/research/
carbon-inequality-2030

Oxfam (2022). Inequality kills. Oxfam GB and Oxfam International, ISBN 978-1-
78748-846-5, https://policy-practice.oxfam.org/resources/inequality-kills-the-
unparalleled-action-needed-to-combat-unprecedented-inequal-621341/

Pain S. (2017). Power through the ages. Nature Outlook, https://www.nature.com/
articles/d41586-017-07506-z

Parravicini V., Svardal K. and Krampe J. (2016). Greenhouse gas emissions from
wastewater treatment plants. *Energy Proceedings*, **97**, 246–253, https://doi.
org/10.1016/j.egypro.2016.10.067

Pate R., Hightower M., Cameron C. and Einfeld W. (2007). Overview of energy-water
interdependencies and the emerging energy demands on water resources. Report
SAND 2007-1349C, Sandia National Laboratories, Albuquerque, New Mexico.
https://www.researchgate.net/publication/228387106_Overview_of_Energy-
Water_Interdependencies_and_the_emerging_energy_demands_on_Water_
Resources

Piani G. (2008a). Il Protocollo di Kyoto: adempimenti, strategie, soluzioni [The Kyoto
Protocol: Compliance, Strategies, Solutions], 720 pp., Zanichelli, Bologna, 2008.
ISBN 978-8808-16648-7

Piani G. (2008b). The unworkable protocol. Nuova Energia (Journal), Milan, Nr.4/2008,
45–51.

Pielke R. (2007). The Honest Broker: Making Sense of Science in Policy and Politics.
Cambridge University Press, Cambridge, UK and New York, USA.

Pierrehumbert R. (2019). There is no plan B for dealing with the climate crisis. *Bulletin
of the Atomic Scientists*, **75**(5), 215–221, https://doi.org/10.1080/00963402.2019.1
654255

Pollan M. (2007). The Omnivore's Dilemma. A Natural History of Four Meals. Penguin
Books, New York.

Poore J. and Nemecek T. (2018). Reducing food's environmental impacts through
producers and consumers. *Science*, **360**(6392), 987–992, https://doi.org/10.1126/
science.aaq0216. Erratum added in Poore J. and Nemecek T. (2021). Science 363,
Issue 6429, https://doi.org/10.1126/science.aaw9908

Raju E., Boyd E. and Otto F. (2022). Stop blaming the climate for disasters. *Communications
Earth & Environment*, **3**, 1, https://doi.org/10.1038/s43247-021-00332-2

Raworth K. (2017). Doughnut Economics. Seven Ways to Think Like A 21ˢᵗ Century
Economist. Penguin, Random House, UK.

Raymond C., Matthews T. and Horton R. M. (2020). The emergence of heat and
humidity too severe for human tolerance. *Science Advances*, **6**(19), 1–8, https://doi.
org/10.1126/sciadv.aaw1838

Realmonte G., Drouet L., Gambhir A., Glynn J., Hawkes A., Köberle A. C. and
Tavoni M. (2019). An inter-model assessment of the role of direct air capture in deep
mitigation pathways. *Nature Communications*, **10**, 3277, https://doi.org/10.1038/
s41467-019-10842-5

REN21 (2020). Renewables 2020, Global Status Report. Ren21 Secretariat, UN Environment Programme, Paris France. https://www.ren21.net/reports/global-status-report/

Revesz R. L., Howard P. H., Arrow K., Goulder L. H., Kopp R. E., Livermore M. A., Oppenheimer M. and Sterner T. (2014). Global warming: improve economic models of climate change. *Nature*, **508**, 173–175, https://doi.org/10.1038/508173a

Richter B. and Boltz F. (eds.) (2020). Building Resilience through Water. Science Direct, Water Security, https://www.sciencedirect.com/journal/water-security/special-issue/10X8CGJB574

Rittel H. W. J. and Webber M. M. (1973). Dilemmas in a general theory of planning. *Policy Sciences*, **4**(2), 155–169, https://doi.org/10.1007/BF01405730

Robine J.-M., Cheung S. L. K., Le Roy S., Van Oyen H., Griffiths C., Michel J.-P., and Herrmann F. R. (2008). Death toll exceeded 70,000 in Europe during the summer of 2003. *National Library of Medicine C R Biology*, **331**(2), 171–178.

Rockström J., Steffen W., Noone K., Persson Å., Chapin F. S. III, Lambin E., Lenton T. M., Scheffer M., Folke C., Schellnhuber H., Nykvist B., De Wit C. A., Hughes T., van der Leeuw S., Rodhe H., Sörlin S., Snyder P. K., Costanza R., Svedin U., Falkenmark M., Karlberg L., Corell R. W., Fabry V. J., Hansen J., Walker B., Liverman D., Richardson K., Crutzen P. and Foley J. (2009). Planetary boundaries: exploring the safe operating space for humanity. *Ecology and Society*, **14**(2), 32, https://doi.org/10.5751/ES-03180-140232

Rodionova M. V., Poudyal R. S., Tiwari I., Voloshin, R. A., Zharmukhamedov S. K., Nam H. G., Zayadan B. K., Bruce B. D., Hou H. J. M. and Allakhverdiev S. I. (2016). Biofuel production: challenges and opportunities. *International Journal of Hydrogen Energy*, **42**(12), 1–16.

Sanders K. T. and Webber M. E. (2012). Evaluating the energy consumed for water use in the United States. *Environmental Research Letters*, **7**, 034034, https://doi.org/10.1088/1748-9326/7/3/034034

Savory A. (2016). Holistic Management. A Commonsense Revolution to Restore our Environment, 3rd edn. Island Press.

Schumacher E. F. (1973). Small Is Beautiful: A Study of Economics as if People Mattered. Random House, UK, 2011.

SDSN/FEEM (2021). Roadmap to 2050: The Land-Water-Energy Nexus of Biofuels. Sustainable Development Solutions Network (SDSN) and Fondazione Eni Enrico Mattei (FEEM), New York. https://www.Roadmap2050.report

SEI, IISD, ODI, E3G, UNEP (2021). The Production Gap Report 2021. http://productiongap.org/2021report

Shiva Kumar S. and Himabindu V. (2019). Hydrogen production by PEM water electrolysis – a review. *Materials Science for Energy Technologies*, **2**(3), 442–454. https://www.sciencedirect.com/science/article/pii/S2589299119300035 https://doi.org/10.1016/j.mset.2019.03.002

Simpson N. P., Mach K. J., Constable A., Hess J., Hogarth R., Howden M., Lawrence J., Lempert R. J., Muccione V., Macheky B., New M. G., O'Neill B., Otto F., Pörtner H. O., Reisinger A., Roberts D., Schmidt D. N., Seneviratne S., Strongin S., van Aalst M., Totin E. and Trisos C. H. (2021). A framework for complex change risk assessment. *Perspective*, **4**(4), 489–501.

Sippel S., Meinshausen N., Fischer E. M., Székely E. and Knutti R. (2020). Climate change now detectable from any single day of weather at global scale. *Nature Climate Change*, **10**, 35–41, https://www.nature.com/nclimate/https://doi.org/10.1038/s41558-019-0666-7

Six J., Paustian E., Elliot T. and Combrink C. (2000). Soil structure and organic matter. Distribution of aggregate-size classes and aggregate-associated carbon. *Soil Science*

Society of America Journal, **64**(2), 681–689, https://acsess.onlinelibrary.wiley.com/doi/epdf/10.2136/sssaj2000.642681x https://doi.org/10.2136/sssaj2000.642681x

Skyllas-Kazacos M. and McCann J. F. (2015). Vanadium redox flow-batteries (VRBs) for medium- and large-scale energy storage. In: Advances in Batteries for Medium and Large-Scale Energy Storage, C. Menictas, M. Skyllas-Kazacos and T. M. Lim (eds.), Elsevier Ltd., pp. 329–386.

Smil V. (2004). Enriching the Earth: Fritz Haber, Carl Bosch, and the Transformation of World Food Production. MIT Press, Washington DC, USA. ISBN: 9780262194495.

Stern N. (2006). The Economics of Climate Change. The Stern Review. Her Majesty's Treasury of the UK Government. Published by Cambridge University Press, London, 2007.

Stern N., Stiglitz J. E. and Taylor C. (2022). The Economics of Immense Risk, Urgent Action and Radical Change: Towards New Approaches to the Economics of Climate Change. National Bureau of Economic Research, Cambridge, Massachusetts, U.S., Working Paper 28472, http://www.nber.org/papers/w28472

Stiglitz J. E. (2013). The Price of Inequality. Penguin, London.

Stone C. (2010). Should Trees Have Standing? Oxford University Press (originally published in 1972).

Tanaka K. R. and Van Houtan K. S. (2022). The recent normalization of historical marine heat extremes. *PLoS Climate*, **1**(2), e0000007, https://doi.org/10.1371/journal.pclm.0000007

Tellman B., Sullivan J. A., Kuhn C., Kettner, A. J., Doyle C. S., Brakenridge G. R., Erickson T. A. and Slayback D. A. (2021). Satellite imaging reveals increased proportion of population exposed to floods. *Nature*, **596**, 80–86, https://doi.org/10.1038/s41586-021-03695-w

UNCTAD (2020). Commodities at a glance. Special issue on strategic battery raw materials. UN Conference on Trade and Development. https://unctad.org/system/files/official-document/ditccom2019d5_en.pdf

UNDRR (2020). Human Cost of Disasters. UN Office for Disaster Risk Reduction. www.undrr.org

UNEP (2008). UN Environment Programme. Vital water graphics. An overview of the state of the world's fresh and marine waters. Increasing price with volume. https://www.unep.org/resources/report/vital-water-graphics-overview-state-worlds-fresh-and-marine-waters

UNEP (2011). Environmental Assessment of Ogoniland. United Nations Environment Program, Nairobi, Kenya. https://www.unep.org/explore-topics/disasters-conflicts/where-we-work/nigeria/

UNEP (2021). United Nations Environment Programme, Emissions Gap Report 2021: The Heat Is On – A World of Climate Promises Not Yet Delivered, Nairobi.

UNEP (2022). Worldwide Food Waste. United Nations Environment Program, Nairobi, Kenya. https://www.unep.org/thinkeatsave/get-informed/worldwide-food-waste

UNIDO (2014). Industrial energy efficiency project. Benchmarking report for the fertilizer sector. United Nations Industrial Development Organization. Prepared by the Austrian Energy Agency and F. Soliman. https://www.unido.org/sites/default/files/files/2019-05/Benchmarking%20Report%20Fertilizer.pdf

UNU-EHS (2021). In Interconnected Disaster Risks, J. O'Connor, C. Eberle, D. Cotti, M. Hagenlocher, J. Hassel, S. Janzen, L. Narvaez, A. Newsom, A. Ortiz Vargas, S. Schutze, Z. Sebesvari, D. Sett and Y. Walz (eds.), United Nations University – Institute for Environment and Human Security (UNU-EHS), Bonn, Germany. https://interconnectedrisks.org/

UN WWDR (2014). Water and Energy. The United Nations World Water Development Report 2014 (2 volumes). United Nations World Water Assessment Programme, Unesco, Paris. http://unesdoc.unesco.org/images/0022/002257/225741E.pdf

UN WWDR (2018). The United Nations World Water Development Report 2018. United Nations Educational, Scientific and Cultural Organization, New York, United States, www.unwater.org/publications/world-water-development-report-2018/

UN WWDR (2020). World Water Development Report 2020 – Water and Climate change. https://www.unwater.org/publications/world-water-development-report-2020/

USEPA (2017). International climate impacts. United States Environmental Protection Agency, 'snapshot' Jan. 19, 2017, the day of the inauguration of President Donald Trump. https://19january2017snapshot.epa.gov/climate-impacts/international-climate-impacts_html (Website information: 'This is not the current EPA website. To navigate to the current EPA website, please go to www.epa.gov. This website is historical material reflecting the EPA website as it existed on January 19, 2017. This website is no longer updated.)

Van Brusselen D., Kayembe-Kitenge T., Mbuyi-Musanzayi S., Lubala Kasole T., Kabamba Ngombe L., Musa Obadia P., Kyanika wa Mukoma D., van Herck K., Avonts D., Devrient K., Smolders E., Banza Lubaba Nkulu C. and Nemery B. (2020). Metal mining and birth defects: A case-control study in Lumumbashi, Democratic Republic of Congo. *The Lancet Planetary Health*, **4**(4), E158–E167, www.thelancet. com/journals/lanplh/article/PIIS2542-5196(20)30059-0/fulltext https://doi.org /10.1016/S2542-5196(20)30059-0

van Rensburg P. (2016). Overcoming global water reuse barriers: The Windhoek experience. *International Journal of Water Resources Development*, **32**, 622–636, https://doi.org/10.1080/07900627.2015.1129319

Vicedo-Cabrera A. M., Scovronick N., Sera F., *et al.* (2021). The burden of heat-related mortality attributable to recent human-induced climate change. *Nature Climate Change*, **11**, 492–500, https://doi.org/10.1038/s41558-021-01058-x

von Bremen A.-H. and Rundgren G. (2020). Kornas planet: om jordens och mångfaldens beskyddare. Ordfront, Sweden, **ISBN**: 9789177750970 (in Swedish, The planet of the cows: about the patron of the Earth and the biodiversity).

WCED (1987). Our Common Future. World Commission on Environment and Development, Oxford University Press, Oxford.

Wiener N. (1948). Cybernetics. MIT Press, Cambridge, Mass., USA (reprint 1961).

WMO (2021a). State of Climate in 2021: Extreme events and major impacts. https://public.wmo.int/en/media/press-release/state-of-climate-2021-extreme -events-and-major-impacts

WMO (2021b). 2021 state of climate services – water. World Meteorological Organization, No 1278, https://public.wmo.int/en/our-mandate/climate/state-of-climate-services-report

Woda J., Wen T., Lemon J., Marcon V., Keeports C. M., Zelt F., Steffy L. Y. and Brantley S. L. (2020). Methane concentrations in streams reveal gas leak discharges in regions of oil, gas, and coal development. *Science of the Total Environment*, **737**, 140105, https://doi.org/10.1016/j.scitotenv.2020.140105

World Bank (2021a). Ebb and Flow: Water, Migration, and Development. https:// www.worldbank.org/en/topic/water/publication/ebb-and-flow-water-migration-and-development

World Bank (2021b). State and Trends of Carbon Pricing 2021. World Bank. © World Bank, Washington, DC, https://openknowledge.worldbank.org/handle/10986/35620 License: CC BY 3.0 IGO

World Economic Forum (2020). The Global Risks Report 2020, 15th edn. www.weforum. org

WRI (2021a). World Resource Institute. Ensuring Prosperity in a Water-stressed World, https://www.wri.org/water

WRI (2021b). World Resource Institute. Global Forest Watch. https://www.wri.org/initiatives/global-forest-watch

Xu R., Yu P., Abramson M. J., Johnston F. H., Samet J. M., Bell M. L., Haines A., Ebi K. L., Li S. and Guo Y. (2020). Wildfires, global climate change, and human health. *The New England Journal of Medicine*, **383**, 2173–2181, https://doi.org/10.1056/NEJMsr2028985

Zabbey N. and Olsson G. (2017). Conflicts – oil exploration and water, Global Challenges, Special issue on Water and energy, http://onlinelibrary.wiley.com/doi/10.1002/gch2.201600015/full

Zabel F., Delzeit R., Scheneider J. M., Seppelt R., Mauser W. and Václavik T. (2019). Global impacts of future cropland expansion and intensification on agricultural markets and biodiversity. *Nature Communications*, **10**, 2844. https://www.nature.com/articles/s41467-019-10775-z https://doi.org/10.1038/s41467-019-10775-z

Zhu W., Jia S., Lall U., Cao Q. and Mahmood R. (2019). Relative contribution of climate variability and human activities on the water loss of the Chari/Logone River discharge into Lake Chad: A conceptual and statistical approach. *Journal of Hydrology*, **569**, 519–531, ISSN 0022-1694, https://doi.org/10.1016/j.jhydrol.2018.12.015

Index

IWA Publishing's authorised EU representative for General Product Safety Regulations is Diane D'Arras, 15 rue Duret, 75116 Paris, France, e-mail: safety@iwap.co.uk.

Printed and bound by CPI Group (UK) Ltd, Croydon, CR0 4YY

12/05/2026

02108891-0004